Motion and Gesture Sensing with Radar

For a complete listing of titles in the
Artech House Radar Library,
turn to the back of this book.

Motion and Gesture Sensing with Radar

Jian Wang
Jaime Lien

ARTECH HOUSE
BOSTON | LONDON
artechhouse.com

Library of Congress Cataloging-in-Publication Data
A catalog record for this book is available from the U.S. Library of Congress.

British Library Cataloguing in Publication Data
A catalogue record for this book is available from the British Library.

Cover design by Andy Meaden Creative

ISBN 13: 978-1-63081-823-4

685 Canton Street
Norwood, MA 02062

10 9 8 7 6 5 4 3 2 1

To my wife, Xiaoli; my children, Grace, Karlos, and Victoria;
as well as my mom, Yunfeng; and in memory of my father, Chenghong Wang

—Jian Wang

For Chris and Theo

—Jaime Lien

Contents

Preface

Radar has widely been considered a mature technology after more than a century of development for defense and aerospace applications. However, recent advances in chip and packaging are now enabling a new radar revolution in the consumer field. The resulting form factor and cost make it possible to embed radar into everyday devices for the first time, opening the possibility for a myriad of new consumer applications.

This modern radar renaissance calls for a fresh look at radar theory and design, directed specifically toward the requirements and constraints of consumer technology. This book presents a complete overview of radar system theory and design for consumer applications, from basic short range radar principles to integration into real-world products. Each chapter aims to provide the reader with an understanding of fundamental theory as well as design procedures, analysis tools, and design examples of radar systems. The book thus serves as a practical guide for engineers and students to design their own radar for motion sensing, gesture controls, and beyond.

Chapters 2–6 of this book cover radar hardware, waveforms/modulations, signal processing, and detection. Chapter 7 discusses machine learning theory and techniques, including deep learning. These techniques are being increasingly applied to radar data and have played a key role in expanding the capabilities of consumer radar. We sincerely thank Dr. Nicholas Gillian for authoring this chapter. Finally, user experience design is an important and integral part of any human interaction sensor. Chapter 8 covers UX design principles and guidelines for radar-based interaction sensing in consumer devices. We express our great appreciation to interaction design experts Dr. Leonardo Giusti, Lauren Bedal, Carsten Schwesig, and Dr. Ivan Poupyrev for contributing this chapter.

Writing this book was a laborious process accompanied by busy day-to-day life and work and compounded by the stressful circumstances of a global pandemic. We greatly appreciate the patience, encouragement, and guidance from our editors, Natalie McGregor and David Michelson.

We also appreciate the valuable suggestions from reviewer Dr. Avik Santra, who truly helped to improve the quality of our book.

Finally, Jian Wang would like to express his deep gratitude to his great friend, colleague, and mentor Dr. Eli Brookner. Dr. Brookner's encouragement and in-depth discussions helped inspire this book. Many of the references are books gifted to me by him. I deeply feel the loss of a friend whom I used to call and have long discussions with on radar, MIMO, and more.

1

Introduction

Radar (radio detection and ranging) is a sensing technology based on the radiation, reflection, and reception of electromagnetic (EM) waves. Radar operates by transmitting radio frequency (RF) waves, which propagate through the atmosphere. Upon encountering changes in the propagation medium (e.g., due to an object, person, or precipitation), some portion of the wave's energy may be scattered and reflected back toward the radar receiver. By processing the reflected waves, or echoes, a radar system can detect, locate, track, characterize, or image the scattering objects or environment. A radar sensor can thus impart *ambient awareness* to a device or system, allowing it to be cognizant of its surroundings and context.

Radar has several attractive and unique features as a contactless sensing modality, particularly for emerging consumer devices. The most prominent attribute of radar is its capability to measure distance and velocity with high accuracy in all weather and lighting conditions. In addition, radars can be easily hidden behind a cover or enclosure due to the ability of RF waves to penetrate materials such as fabric, plastics, and glass. This is particularly important for product and industrial design due to aesthetic considerations, protection from dust and moisture, and integration flexibility.

As a technology invented more than 100 years ago, the fundamental principles of radar are mature. The concept was first demonstrated by Hertz in the late nineteenth century [1], and further developed by other pioneer scientists such as Hulsmeyer and Marconi [2]. In ensuing years, radar was forgotten and rediscovered a few times before finding critical usage during the World Wars as a system for detecting aircraft. Since then, radar has been widely adopted for numerous military, air traffic control (ATC), remote sensing, navigation, space, and civil applications, playing a key role in many safety critical missions. Today, the U.S. airspace is seamlessly covered by the Common Air Route Surveillance Radar (CARSR) network [3]. Similar radar systems are also deployed in many other countries. Modern aircraft rely on onboard weather radar to avoid hazardous storms and turbulence. The Next Generation Weather Radar (NEXRAD) system [4] is a network of 160 radars covering most of the North American continent, providing near real time detection of precipitation and wind.

Several recent technological advancements have contributed to an emerging renaissance for radar beyond military and civil use and into the consumer space. In particular, new silicon technologies at higher frequencies [5] are making it possible to drastically shrink the size, weight, and power consumption of radar. State-of-the-art silicon-germanium (SiGe) and complementary metal oxide semiconductor (CMOS) processes have enabled full radar system-on-chip implementations, including antennas on package, that can be manufactured at scale and easily integrated into consumer devices. Compared to a traditional airport surveillance radar that costs multimillions of dollars, consumes kilowatts of power, and needs a few 42U cabinets to host its electronics, a complete radar-on-chip today has a footprint on the order of square millimeters, power consumption in the order of milliwatts, and cost of a few dollars. The radar technology landscape is thus rapidly transforming from cumbersome discrete systems to consumer pico-sensors with massive market volume.

These technology developments have opened the door for radar to be integrated into numerous consumer products such as cell phones, smart displays, and watches. Several consumer products with integrated radar pico-sensors have already hit the market, bringing with them new interaction paradigms and use cases for touchless human-machine interfaces [6]. The last decade has witnessed a drastic increase in research and development for radar technology and applications, as well as an explosion of new companies and products incorporating radar for key features. With an open field for exploration across all aspects of modern radar, including hardware, software, algorithms, machine learning, interaction design, and use cases, we expect that this radar renaissance is just beginning and will continue for years to come.

1.1 Radar Basics and Types

The basic principles of radar are straightforward, as illustrated in Figure 1.1. The transmit system generates an EM waveform at a specific carrier frequency and bandwidth with desired modulations. The waveform can be a simple pulse of sinewave or modulated in amplitude, phase, or frequency. The waveform is radiated through a transit antenna into the space. Some portion of the EM energy is intercepted by the target in the field of view and scattered in many directions. Some of these scattered signals are received by the receive antenna

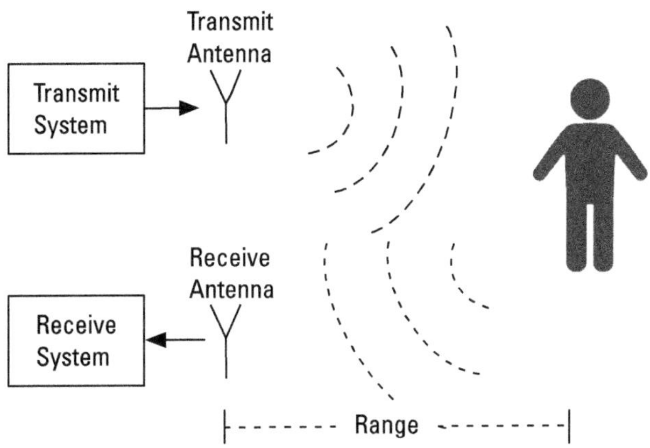

Figure 1.1 A rudimentary radar system.

and receive system, where it is processed to provide the measurement information including the presence and location of the target. The target range is obtained through measuring the travel time passing between the transmission of the radar waveform and the reception of its echo back from the target. This travel time t is also called time of flight and the corresponding target range R (also known as slant range) is:

$$R = ct / 2 \tag{1.1}$$

where c is the speed of light and factor of 2 is due to two-way propagation of the radar signals. When there is relative motion between the radar and target, there is a frequency shift in the return signal compared to the transmit one due to the Doppler effects (detailed discussion in Chapter 3). This shift is proportional to the radial velocity of the target relative to the radar and is often used to improve target detection among clutter (unwanted returns from the environment such as ground, ocean, and rain). The target's angle information can be measured through a scanning narrow-beam receive antenna or receive antenna array (the processing is discussed in Chapter 5).

When the transmit and receive systems are collocated, this type of radar is called monostatic radar; otherwise, it is known as bistatic system. Most of civil radars are monostatic. Radars can also be classified based on their operating waveforms: pulsed radar or continuous wave (CW) radar. Pulsed radar transmits pulse waveform as an example shown in Figure 1.2. There is generally more than one pulse in the processing window. These pulses repeat at a pulse repetition frequency (PRF) that determines the maximum unambiguous range and Doppler frequency. The inverse of PRF is the pulse repetition interval (PRI). In

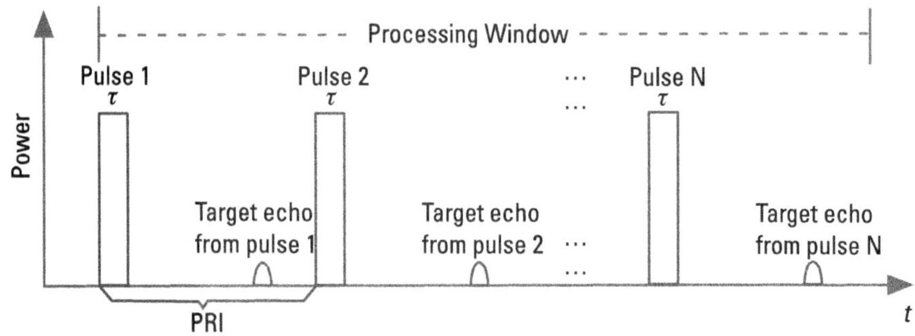

Figure 1.2 Example of pulse waveform. The rectangular pulses represent short duration (τ) sinewaves, and the target returns are at much lower power.

conventional pulsed radar, the power of the transmitted pulses is high, and the receiver needs to be gated off during the transmission to protect its electronic components. Under such circumstances, the pulse duration τ determines the minimum detection range:

$$R_{min} = \tau c / 2 \tag{1.2}$$

where τ is the duration of the pulse. These pulsed radars cannot detect targets within the minimum range.

In CW radar, the receiver is on while the transmitter is operating. CW radar can detect very close in targets with minimum range of zero. If the transmitter power is more than the receiver can tolerate, CW radar can be in bistatic configuration to increase the transmitter to receiver isolation.

In recent development of civil applications, the boundary between CW radar and pulsed radar is blurred. In these applications, the targets of interest are from very close range to only a few meters or a few hundred meters. This posts significantly lower (>100 dB) transmit power requirement compared to that of conventional surveillance radar designed to detect targets as far as a few hundred nautical miles. Because of the low transmit power and desire to have a minimum detection range of zero, these pulsed civil radars keep their receivers on even during the pulse waveform transmission. On the other hand, CW radars, especially frequency-modulated continuous wave (FMCW) radar, introduce gaps between two neighboring chirps to save energy and computational resources. These FMCW radars have a similar waveform structure to a pulse train. Their chirp repetition interval (CRI) is equivalent to PRI, and their signal processing is the same as that of pulsed radar after demodulation. The detailed discussion can be found in Chapter 4. In this book we will simply use waveform repetition interval (WRI) when discussing either pulse train's PRI or chirp train's CRI.

1.2 Frequency Bands and Civil Applications

Radar generally operates in the so-called microwave frequency range. This is not a strictly defined range, and operational radars can be found at any frequency from a few megahertz to terahertz. The IEEE has adopted a letter-band system [7] as a standard, as shown in Table 1.1. These letters were originally developed to keep military secrecy and later accepted and used by radar and communication engineers.

HF Band

The main civil applications of HF radar are 200 nautical mile (nmi) exclusive economic zone (EEZ) monitoring [8] and ocean currents mapping [9]. In HF band, the EM wave can couple with the surface of salty water and propagate

Table 1.1
Standard Radar-Frequency Letter-Band Nomenclature and Application

Band Designation	Nominal Frequency Range	General Applications
HF	3–30 MHz	Over the horizon radar; very long range (up to 3–4 thousand kilometers) with low spatial resolution and accuracy (in the order of kilometers)
VHF	30–300 MHz	Similar to UHF band
UHF	300–1,000 MHz	Long range surveillance and early warning (up to 500 km); weather effects are negligible; low to medium spatial resolution and accuracy
L	1–2 GHz	Long range surveillance (up to 370 km) with medium spatial resolution and accuracy; en-route air traffic control and weather monitoring
S	2–4 GHz	Medium range surveillance (~100 km); severe weather can limit the detection range; weather radar network (Nexrad)
C	4–8 GHz	Short range surveillance; increased spatial resolution and accuracy; subject to significant weather effects in heavy rain; ultrawideband (UWB) radar
X	8–12 GHz	Short range surveillance; airplane on board weather avoidance; increased effects from weather; increased resolution and accuracy
K_u	12–18 GHz	Short range tracking and guidance; satellite-based radar system; police radar for overspeed detection
K	18–27 GHz	
K_a	27–40 GHz	
V	40–75 GHz	Very short range detection and tracking; excellent resolution and accuracy due to the 7-GHz bandwidth at 60 GHz; new applications on human motion detection and gesture control
W	75–110 GHz	Very short range detection and tracking; good resolution and accuracy; automotive radar
mm	110–300 GHz	Very short range tracking and guidance

beyond the horizon in ground wave mode. The remote measurement of ocean surface currents is done through exploiting a Bragg resonant backscatter phenomenon. Networks of HF radar systems are capable of monitoring surface currents up to 200 km with a horizontal resolution of a few kilometers at an hourly basis.

L and S Bands

The main civil applications of this band are air traffic control and weather detection and monitoring networks. The CARSR network in L band covers U.S. regions seamlessly, which makes flying safer and is key to tracking hostile aircraft and preventing terrorist attacks. If such a radar network existed in Malaysia, the missing MH 370 airplane would have been found right away. The Nexrad weather radar network is working at S band and probably is the first and largest in the world. Other developed countries also followed to form their own weather monitoring systems at either S band or C band.

X Band

Radar at X band is small enough to be mounted in the nose of commercial aircraft. Even with increased attenuation loss, radar at this band can still detect turbulence up to 40 nmi and heavy weather up to 320 nmi [10]. The processed 3-D weather displays are provided to the pilots in real time to help enhance flight paths based on weather patterns. The pilot can make small maneuvers and avoid hazardous or turbulent routes to ensure the safety of passengers and flight crews. This also significantly improves the flight comfort of the passengers. Nowadays most airlines and business jets are equipped with on-board weather radar.

V Band

The most active research and development activities in V band are probably within 60 GHz. There is a continuous 7 GHz (57–64 GHz) bandwidth for high data throughput. At 60 GHz, oxygen molecules in the atmosphere resonate with the RF signals to cause much larger attenuation compared to that of neighboring frequencies. This characteristic makes it an ideal candidate for crosslink communication between satellites in a constellation with protection against interception by ground-based stations. The land-based communication application is the WiGig (60 GHz Wi-Fi) standard (IEEE 802.11ad and IEEE 802.11ay) with data transfer rates up to 7 Gbps for very short ranges up to ~10m.

In addition, 60 GHz is permitted for radar applications globally. The Federal Communications Commission (FCC) has strict rules (FCC 15.255) that require that the peak transmitter conducted output power shall not exceed –10 dBm and the peak EIRP level shall not exceed 10 dBm. The FCC granted

Google a waiver in 2018 [11], allowing a peak transmitter conducted output power of +10 dBm and a peak EIRP level of +13 dBm to support its Soli radar's gesture-recognition and motion sensing [12, 13]. In this waiver there is also a duty cycle limit of 10% in any 33-ms interval. Following Google's waiver there have been a few similar ones granted, such as the in-cabin automotive motion sensors used to monitor for children left behind and collision-avoidance systems for specialized drone operations. The interests in this band are so high that FCC decided to have a rulemaking in 2021 to formally change the rule to allow higher transmit power. The rule change is likely to happen in 2022 and is going to open the door for more technological uses and innovations in the 57–64 GHz band.

A few new products based on 60-GHz radar have been launched, and many are under development. The first mass market release is the Pixel 4 phone, which is also the first cell phone in history to integrate a complete radar system under the display bezel, as shown in Figure 1.3. The radar chip including antenna in a package is the size of 5 × 6.5 mm, and the evolution of the chip size is shown in Figure 1.4. This tiny chip enables the cell phone to sense people nearby and provides contactless gesture control. Soli radar was later incorporated into Google's Nest Thermostat and Hub products.

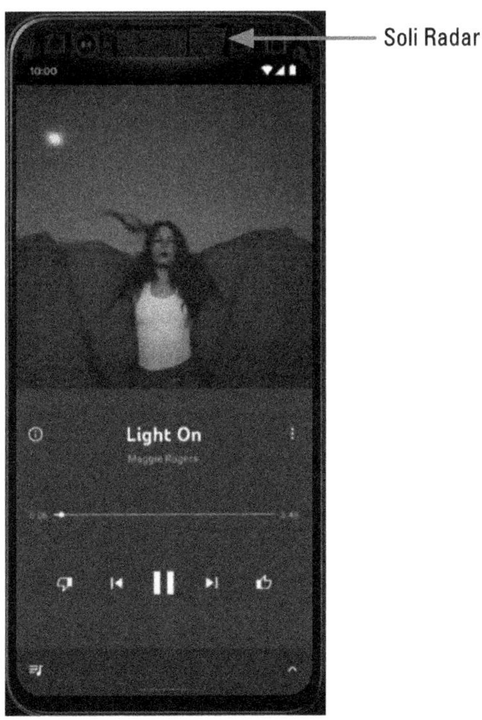

Figure 1.3 Soli radar is inside the Pixel 4 phone.

Figure 1.4 Size evolution of Soli chips to be able to fit into a cell phone.

W Band

The 76–81-GHz band has been assigned to automotive radar to replace the 24-GHz band, currently being phased out. As the consumers increasingly value the advanced driver assist systems (ADAS), automotive radar as a core component of ADAS has reached a multibillion dollar market size. This market is expected to grow at 10% compound annual growth rate in the next decade or so [14]. There are two types of automotive radars [15]: the wideband high-precision short-range vehicular radar (SRR) and the long-range vehicular radar (LRR). The SRR is mainly used for blind spot detection and autonomous emergency braking; the LRR is mainly used for adaptive cruise control. The working principle and signal processing of automotive radar are very similar to those of motion detection and gesture control radars.

1.3 Radar Standardization

Radar as a sensor technology has enormous applications in the big field of Internet of Things (IoT) and smart devices. As an effort to help accelerate these application developments, Ripple [16] (hosted by CTA and initiated by Google) is developing an open-source API standard to enable interoperability and growth of applications. Development of applications for radar is time consuming and requires domain knowledge, and interoperable SW libraries for the radar simplify the application development process.

1.4 Book Outline

Radar was considered a mature technology after more than a century of development for defense and aerospace applications. However, recent advances in millimeter-wave technology are now enabling a new radar revolution in the consumer field, driven by shrinking chip sizes and more scalable silicon manufacturing processes. The resulting form factor, cost, and power make it possible to embed radar into everyday devices and open the possibility for a myriad of new consumer applications.

This modern radar renaissance calls for a fresh look at radar theory and design, directed specifically toward the requirements and constraints of consumer technology. This book provides a complete overview of radar system theory and design for consumer applications, from basic short range radar theory to the integration into the real-world products. It provides the theoretical understanding, design procedures, analysis tools, and design examples of radar systems. The book provides practical guidance for engineers and students to design their own radar in consumer electronics for motion sensing and gesture controls. The book is self-contained to cover radar hardware, waveforms and demodulation, signal and array signal processing, detection and classification, machine learning, and UX design.

References

[1] Hertz, H., and D. E. Jones, *Electric Waves*, Book on Demand Ltd., 2013. (Replication of a book originally published before 1893.)

[2] Marconi, S. G., "Radio Telegraphy," *Proc. IRE*, Vol. 50, No. 8, 1962, pp. 1748–1757.

[3] Wang, J., et al., "Modernization of En Route Air Surveillance Radar," *IEEE Transactions on Aerospace and Electronic Systems*, Vol. 48, No. 1, January 2012, pp. 103–115.

[4] Crum, T. D., and R. L. Alberty, "The WSR-88D and the WSR-88D Operational Support Facility," *Bulletin of the American Meteorological Society*, Vol. 74, No. 9, 1993, pp. 1669–1688.

[5] Saponara, S., et al., *Highly Integrated Low-Power Radars*, Norwood, MA: Artech House, 2014.

[6] Lien, J., et al., "Soli: Ubiquitous Gesture Sensing with Millimeter Wave Radar," *ACM Transactions on Graphics*, Vol. 35, Issue 4, No. 142, 2016, pp. 1–19.

[7] "IEEE Standard Letter Designations for Radar-Frequency Band," IEEE Std 521-1984, 1984, pp. 1–8.

[8] Ponsford, A. M., and J. Wang, "A Review of High Frequency Surface Wave Radar for Detection and Tracking of Ships," *Turkish Journal of Electrical Engineering and Computer Science*, Vol. 18, No. 3, 2010, pp. 409–428.

[9] Paduan, J. D., and L. Washburn, "High-Frequency Radar Observations of Ocean Surface Currents," *Ann Rev Mar Sci.*, Vol. 5, 2013, pp. 115–136.

[10] Challa, H. B., et al., "Review of Weather Radars—Past, Present and the Scope for Future Modifications with Technology Innovation," *International Journal of Advance Research in Science and Engineering*, Vol. 7, No. 2, 2018.

[11] Federal Communications Commission: DA 18-1308 "Matter of Google LLC Request for Waiver of Section 15.255(c)(3) of the Commission's Rules Applicable to Radars used for Short-Range Interactive Motion Sensing in the 57-64 GHz Frequency Band," December 31, 2018.

[12] "Soli," Google Advanced Technology and Projects, https://atap.google.com/soli.

[13] Trotta, S., et al., "2.3 SOLI: A Tiny Device for a New Human Machine Interface," *Proc. 2021 IEEE International Solid-State Circuits Conference*, 2021, pp. 42–44.

[14] "Automotive Radar Market Research Report Summary," *Fortune Business Insights*, January 2022.

[15] Waldschmidt, C., et al., "Automotive Radar—From First Efforts to Future Systems," *IEEE Journal of Microwaves*, Vol. 1, No. 1, 2021, pp. 135–148.

[16] Ripple-Radar Standard API, Consumer Technology Association, https://cta.tech/ripple.

2

Radar System Architecture and Range Equation

The size, weight, power, and cost of radar are critical to civil applications, especially for mobile platforms. Most of these radar systems are working at the 60-GHz band (57–64 GHz) for short-range interactive motion sensor and at 76–81-GHz band for advanced vehicular applications. The wavelength of these bands is in the range of millimeters and these radars are also referred to as mm-wave radars. The vast majority of these radars adopt a linear frequency modulated continuous wave (LFM CW) transceivers architecture due to their simple transmitter structure and low-cost receivers. There are also pulse radar systems, especially when they share hardware components with communication systems such as WiGig (IEEE 802.11 ad & ay) systems. In this chapter we discuss the common architecture of these systems and their hardware components. We also describe how to analyze these systems and predict their performance through radar range equation.

2.1 Basic Hardware Components of Radar

A general radar hardware architecture is shown in Figure 2.1, showing a transmitter, waveform generator, receiver, antenna, and signal data processor. The waveform generator produces a low-power radar signal at the designated carrier frequency, which is amplified in the transmitter to reach the desired power. The output of the transmitter is fed to transmit antennas through a transmission line, where it is radiated to the space. The receiver will down-convert and digitize the received signals from receive antennas. The transmitter and receiver

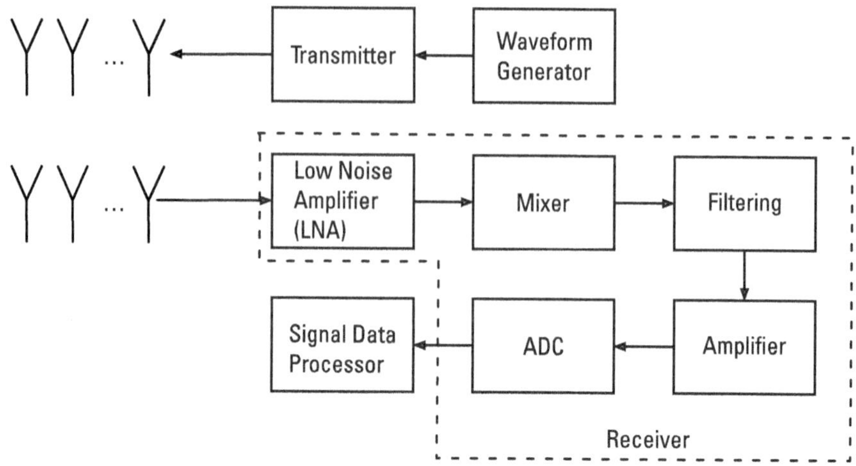

Figure 2.1 Block diagram of a civil radar.

components are generally collocated in civil applications, and this type of radar is referred to as monostatic radar. When the transmitter and receiver are separated by a distance comparable to the expected target distance, the corresponding radar is called bistatic radar. In this book we will only consider monostatic radar.

2.1.1 Transmitter/Receiver (Transceiver)

Conventional radar transmitter usually consists of multistage amplifications to achieve hundreds to millions of watts of output power. However, in civil application the output power is on the order of tens to hundreds of milliwatts, and there is only a single stage solid state amplifier required. The receiver's function is to down-convert the received RF signals to IF or baseband prior to digitizing. The first part of the receiver is generally a low noise amplifier (LNA), which amplifies the return signals to better compete with the noise sources downstream. The LNA itself also generates noise, which dominates the overall system noise level as explained in Section 2.3. This is the reason that it is important to use amplifiers with minimum internally generated noise. LNAs are more expensive and have limited dynamic range (capability to handle returns from both strong and weak targets simultaneously). In many civil radar systems, the received signals are fed directly into a mixer to have larger dynamic range but at the cost of higher system noise level. A typical mixer has >10 dB more dynamic range than that of an LNA. This extra dynamic range is due to the fact that the mixer's compression point is at a much higher input power level. Beyond the compression point, the device starts to saturate and is no longer linear. Under such circumstances, its outputs will have distortions, harmonics, and intermodulation products. Therefore, it is important to design the RF front end to avoid saturation. In many civil applications, the radar coverage starts at nearly zero range, and a high compression point is desirable to handle possible strong returns from close in targets or interference. The tradeoff between dynamic range and system noise level should be determined based on the specific application requirements. In automotive radar, it is more prevalent to have mixers as the first RF stage due to the stringent requirement of dynamic range.

The current process technology for transceivers is migrating from SiGe/BiCMOS to CMOS [1–3] due to the ease of digital circuits integration and low cost of CMOS. There are products available based on both technologies in the market.

P1dB Compression Point

Generally, the analog device, such as an amplifier, is working in linear mode, which means the output power is the input power plus a fixed gain if all in decibel domain. The P1dB compression point is the input power at which the corresponding output power is 1 dB less than that it should be. The P1dB compression point is the property of the analog device and beyond which the device is in nonlinear mode.

Mixer

The mixer is an analog circuit that generates output signals whose frequencies are the sum and difference of the two input signals. One of the input signals is from received radar echoes with carrier frequency f_{RF}, and the other one is from the local oscillator (LO) with frequency f_{LO}. Therefore, the LO is an integral part of the mixer. The output of the mixer is followed by a filter that will let the signal with the difference frequency pass through. The difference frequency is also known as intermediate frequency (IF):

$$f_{IF} = \left| f_{RF} - f_{LO} \right| \tag{2.1}$$

When carrier frequency is identical to the LO frequency, this type of architecture is referred to as direct-conversion or homodyne system, and the IF frequency is zero hertz. The LFM radars are homodyne systems where the RF signals are down-converted to baseband in one stage.

When the two frequencies are different, it is referred to as superheterodyne system with the nonzero f_{IF} chosen for low-noise and low-cost circuits. In this kind of system, special care has to be given to the problem of image frequency. An image frequency is the undesired frequency after mixing equals to the IF but with the opposite sign. Depending on whether f_{RF} is larger or smaller than f_{LO}, the corresponding image frequency is

$$f_{image} = \begin{cases} f_{LO} - f_{IF} = f_{RF} - 2f_{IF} & f_{RF} > f_{LO} \\ f_{LO} + f_{IF} = f_{RF} + 2f_{IF} & f_{RF} < f_{LO} \end{cases} \tag{2.2}$$

After mixing, the noise and interference at image frequency appear at f_{IF} and cannot be separated from the desired signal. It is important to filter the image frequency signals prior to mixing to avoid performance degradation. The filter design is easier if f_{image} and f_{RF} are widely separated. Therefore, there might be multiple mixing stages in superheterodyne systems, and the first stage can utilize a higher IF to allow the image frequency band to be further away from the desired signal band for the ease of filter design.

The simplest form of mixer is just a single diode terminating a transmission line, which is also known as a single-ended mixer. The mixer can produce intermodulation products at other frequencies when the following condition is met:

$$nf_{RF} + mf_{LO} = f_{IF};\ m,n = 0,\pm 1,\pm 2,\ldots \tag{2.3}$$

This type of mixer can also couple the noise of LO into IF band. To address this issue, more complex mixers can be used, such as a balanced mixer [4]. There is also an image rejection mixer [5] if an RF filter is not an option to remove the image frequency.

A/D Converter

The A/D converter (ADC) converts analog signals to digital signals to enable the following digital processing. There are many different types of ADCs, such as successive approximation (SAR), delta-sigma, pipelined, and flash [6]. SAR ADC has a relatively low power, cost, and sampling rate. Pipelined and flash ADCs have relatively high power, cost, and sampling rates, and delta-sigma ADC is in between. When selecting ADC for a specific application, the most important system-level performance indicators are the number of bits and sampling rate. The number of bits into which the signal is quantized inside the ADC is disproportional to the sampling rate (bandwidth). Therefore, the signal with a wide bandwidth has more difficulty maintaining good performance. For example, the flash ADC has a large array of comparators, and the input signal is compared to multiple reference voltages in parallel. This enables the fast-sampling rate of the flash ADCs; however, the complexity of the comparator network also restricts the available number of bits. SAR ADC, on the other hand, uses only a single comparator to sequentially compare the input signal to the reference points. It is easier to have high accuracy (more bits) but at a much lower rate.

When the ADC converts an analog input voltage to a digital value, there is a rounding error, which is also known as the quantization error. Let us assume that the quantization error $e(t)$ is uniformly distributed between $-1/2$ LSB and $+1/2$ LSB, and the input signal is also uniformly distributed to cover all quantization levels (this assumption is not precise, however it is accurate enough for most applications). If we use q to indicate the LSB, the mean square value of $e(t)$ is:

$$\overline{e^2(t)} = \int_{-q/2}^{q/2} \frac{1}{q} x^2(t)\,dt = q^2/12 \tag{2.4}$$

If the ADC has N bits, the maximum input sinusoidal signal without saturating the ADC is:

$$s_{fs}(t) = q2^{N-1}\sin(2\pi ft) \tag{2.5}$$

which has a power of

$$P_{fs} = q^2 2^{2N-3} \tag{2.6}$$

The signal-to-noise ratio (SNR) of the ADC is defined as:

$$SNR_{adc} = P_{fs} / \overline{e^2(t)} = 1.5 \times 2^{2N} \tag{2.7}$$

The logarithm form of (2.7) is:

$$SNR_{adc} = 6.02N + 1.76\,dB \tag{2.8}$$

2.1.2 Waveform Generator

The radar waveform is generated by the waveform generator. There are various different architectures, from the well-known simple pulse generator (for pulse waveforms) and phase locked loop (PLL) (for frequency modulated waveforms) to the more advanced direct digital synthesis (DDS) (for arbitrary waveforms) [7].

The pulse generator is limited to binary phase (0 and 180 degrees) coded pulse train, and the PLL architecture is suitable for simple frequency modulations such as linear FM. They, however, have good performance and low cost, and are widely adopted in civil radar systems.

The DDS is more popular in military and aerospace radar systems. For example, the air traffic control radar implements DDS to produce nonlinear FM waveforms [8]. DDS is more flexible, precise, and agile over analog techniques. A general architecture of an arbitrary waveform generator using DDS is shown in Figure 2.2, where a digital baseband waveform is modulated with digitalized sine and cosine signals generated by a numerically controlled oscillator (NCO) to produce modulated carrier signals. These signals are then digitally summed and converted to analog through a digital-to-analog converter (DAC) before passing through a filter to produce the IF output. If the baseband complex waveform is $e^{j\theta(t)}$, then the IF output can be expressed as:

$$s(t) = \cos(\theta(t))\cos(2\pi f_{IF}t) - \sin(\theta(t))\sin(2\pi f_{IF}t) = \cos(2\pi f_{IF}t + \theta(t)) \quad (2.9)$$

Depending on the form of $\theta(t)$, the IF output can be any phase/frequency modulated signals. The digital waveform generator is not yet popular in civil applications due to its relatively high cost. Quantization and nonlinearity of DAC and spurs due to digital clock and circuits should also be handled with extra care in order to achieve good performance [9].

2.1.3 Antennas

The antenna is effectively a transducer between propagation in space and in transmission lines. Transmit antennas radiate electromagnetic energy in the direction of targets, and receive antennas collect energy scattered back by targets. Very often a common antenna can be used for both transmission and reception. When discussing antenna properties, conclusions obtained for one can be applied to the other due to antenna reciprocity theorem [10]. For example, there is no distinct transmit and receive beampatterns for the same antenna.

There are many types of antennas [11], such as wire (dipole, monopole, and loop), aperture (waveguide and horn), reflectors, lens, and more recent microstrip and arrays. These antennas differ in how the radiation beams are

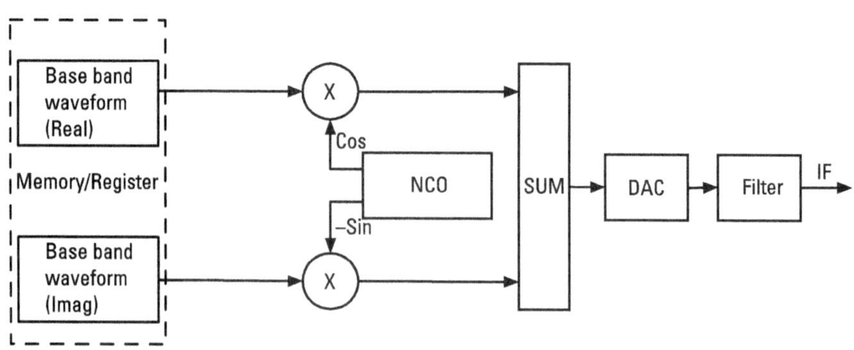

Figure 2.2 Digital waveform generator.

formed and steered in space. For civil applications, the antenna size and integration efforts are of most importance and the low-profile antennas such as microstrip are widely adopted. They are often printed in the printed circuit board (PCB) or in the package [12]. Due to the same reason, the beams are electronically steered through an array of printed antennas to avoid any moving parts.

Directive and Power Gain

Gain is a measure of the antenna's ability to focus energy into a particular direction. The directive gain (also referred to as directivity) describes the fundamental antenna radiation pattern and is often used by antenna engineers. The power gain definition also considers the loss of the antenna and is more appropriate to be used in the radar range equations by radar engineers.

The directive gain is defined as [13]:

$$G_D = \frac{\text{maximum radiation intensity}}{\text{average radiation intensity}} = \frac{\max_{\theta,\phi} P(\theta,\phi)}{\text{total power radiated} / 4\pi} \quad (2.10)$$

where the radiation intensity is the power per unit solid angle radiated at a particular direction (θ, ϕ) and denoted as $P(\theta, \phi)$; the average radiation intensity over the entire solid angle of 4π is equal to the total power radiated divided by 4π.

The power gain takes the dissipative losses of the antenna into account and is related to directive gain as:

$$G = G_D \rho \quad (2.11)$$

where ρ is the radiation efficiency defined as the ratio of the total power radiated to the overall power received by the antenna at its terminals.

For a receive antenna, its gain is also related to its effective aperture A_e [11]:

$$G_r = \frac{4\pi A_e}{\lambda^2} \quad (2.12)$$

where λ is the wavelength, and A_e is a measure of the effective area to receive the incident wave. Equation (2.12) is just an approximation for low loss antenna; however, it is widely used in the radar range equation.

Antenna Radiation/Beam Pattern

When the radiation intensity is normalized with its maximum equal to unity, the plot of this normalized intensity as a function of the angular coordinates is called antenna radiation pattern or simply radiation pattern. It is also known as the antenna beampattern. Figure 2.3 shows an example antenna beampattern in azimuth plane for an eight-element uniform linear array. The main beam is at zero degrees and the rest of the pattern outside the main beam is sidelobes. The width of the main beam, the sidelobe levels, the depth of nulls, and their position are all design parameters and need to be determined based on the particular applications. In many applications, narrow main beam and low sidelobes are desirable for good angle resolution and accuracy. However, for motion and gesture sensing, a wide field of view is more important, and the antenna should be designed to be as close as omnidirectional. This means within the field of view: the main beam should be wide and flat; sidelobe should be high; and the nulls should be as shallow as possible.

For the ease of discussion, Figure 2.3 shows only the beampattern in one dimension, although a complete beampattern should be a function of both azimuth and elevation or other appropriate angle coordinates.

Polarization

The antenna polarization is defined as the orientation of the electric field of the radiated electromagnetic wave. Most radar antennas in civil applications are

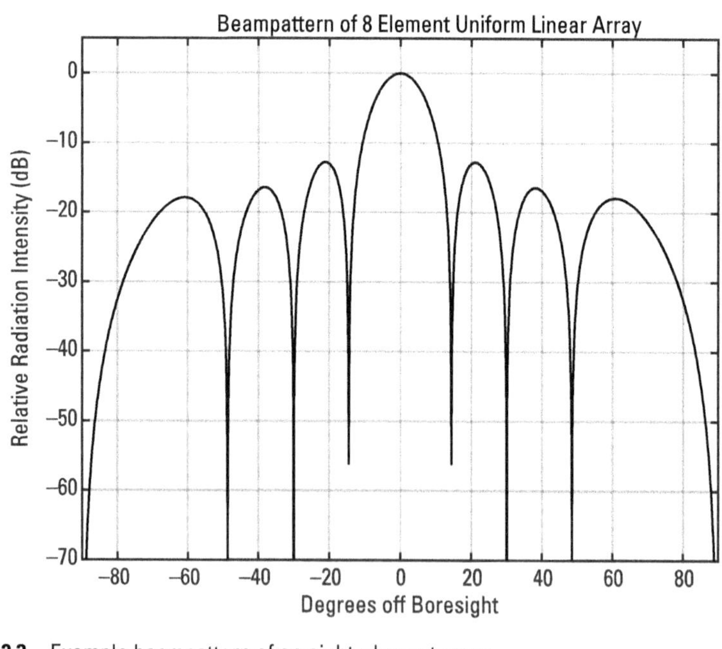

Figure 2.3 Example beampattern of an eight-element array.

linearly polarized with the orientation being either horizontal or vertical. There are generally no particular requirements for one polarization over the other.

2.2 LFM Radar Architecture

The LFM radar is well suited to civil applications such as human and gesture sensing and automotive radar. This is mainly due to the fact that LFM radar can utilize large frequency bandwidth to achieve fine-range resolution while remaining relatively low cost. A general architecture is shown in Figure 2.4, where the LFM signal is generated by a PLL and amplified through a power amplifier (PA). The amplified signal can be subject to simple binary phase coding (BPSK) before being fed to an antenna. On receive, the echo signals are down-converted to base band directly prior to being sampled. The high-pass filter is necessary to remove the leakage and interference near DC, and the low-pass filter is used to limit the bandwidth of the signals and prevent aliasing during the sampling process in ADC.

2.3 Receiver Noise

The sensitivity of radar is limited by noise, and in most cases the internal noise within the receiver as opposed to external noise sources (e.g., atmosphere or sun) dominates the overall noise level. Even with a perfectly designed receiver with no excess noise, thermal noise is still generated by the thermal agitation of the electrons in the ohmic section of the receiver's input circuits. The thermal noise power P_n at the input of the receiver is determined by the noise bandwidth B_n and absolute temperature T of the input circuits [14] regardless of applied voltage:

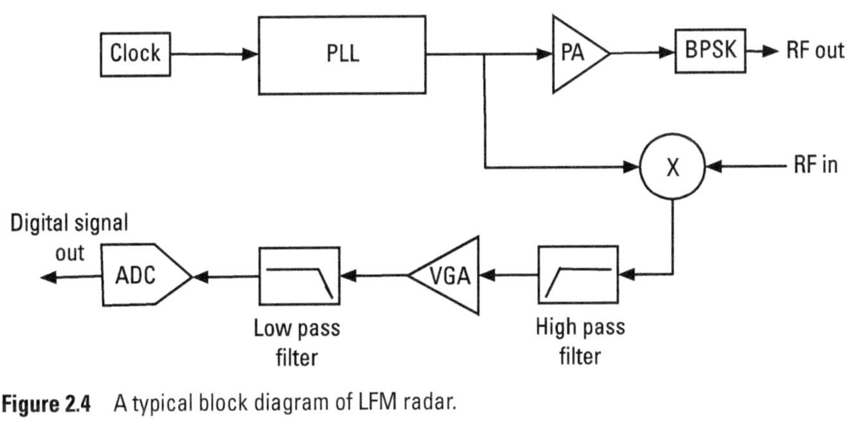

Figure 2.4 A typical block diagram of LFM radar.

$$P_n = kB_nT \tag{2.13}$$

where $k = 1.38 \times 10^{-23}$ J/deg. is the Boltzmann's constant. The noise bandwidth should be defined at the matched filter and is often approximated by its half power (3-dB) bandwidth B in radar system performance analysis. The concepts of matched filter and half power bandwidth are explained in the following chapter. The imaginary concept of noise temperature T_n is defined as:

$$T_n = P_n / kB_n \tag{2.14}$$

The actual noise power in a practical receiver is higher than the thermal noise alone. The measure of the actual noise is through the so-called noise figure, which is defined as:

$$\begin{aligned} F_n &= \frac{\text{actual noise output of practical receiver}}{\text{thermal noise output of ideal receiver at standard temperature } T_0} \\ &= \frac{N_{out}}{kT_0BG_{Rx}} \end{aligned} \tag{2.15}$$

where G_{Rx} is the gain of the receiver, and $T_0 = 290$ K. The receiver gain G_{Rx} is the ratio of output signal power S_{out} to input signal power S_{in}, therefore (2.15) can be expressed as:

$$F_n = \frac{N_{out}S_{in}}{kT_0BS_{out}} = \frac{S_{in} / kT_0B}{S_{out} / N_{out}} = \frac{SNR_{in}}{SNR_{out}} \tag{2.16}$$

where SNR_{in} is the signal-to-noise ratio at the receiver input and SNR_{out} is the signal-to-noise ratio at the receiver output. Equation (2.16) indicates that the noise figure can be considered a measure of the signal-to-noise ratio degradation after the signal passes through the receiver.

There are many components inside a receiver; however, the noise figure is mainly determined by these components up to and including the initial amplification. This can be seen by examining the noise figure of cascade of components. If the first component in the chain has noise figure F_1 and gain G_1, the second component has noise figure F_2 and gain G_2, and so on; the overall noise figure of the chain is [15]:

$$F_n = F_1 + \frac{F_2 - 1}{G_1} + \frac{F_3 - 1}{G_1G_2} + \frac{F_4 - 1}{G_1G_2G_3} + \ldots \tag{2.17}$$

If G_1 is large enough, all the following components' noise contributions can be neglected. That is why in many radar systems the LNA is used as the first amplification component to minimize the overall noise figure.

For a receiver with overall noise figure F_n and gain of G_{Rx}, the input noise is:

$$P_n = kBF_nT_0 \tag{2.18}$$

and the output noise is:

$$P_{no} = G_{Rx}kBF_nT_0 \tag{2.19}$$

Equation (2.18) contains thermal noise and the extra noise generated by receiver, and the noise temperature due to the receiver itself is:

$$T_e = \frac{kBF_nT_0 - kBT_0}{kB} = (F_n - 1)T_0 \tag{2.20}$$

Noise temperature by default is defined at the input of the component. When specified as output noise temperature, the corresponding gain (or loss for passive component) should be applied to its noise temperature. For example, the previous receiver's output noise temperature is $(F_n - 1)\,T_0G_{Rx}$.

2.4 Dynamic Range

Dynamic range is the ratio of maximum input signal power to minimum input signal power, which can be simultaneously handled by the receiver without performance degradation. The maximum input signal power is generally the value that causes the first stage amplifiers (either LNA or mixer) to reach its P1dB saturation point. The minimum signal is often referring to the receiver noise at the input.

A large dynamic range is important in that the targets can be misdetected when the receiver is saturated, and it takes finite time to recover. The dynamic range of the receiver could be limited by the ADC if it does not have enough bits. The dynamic range or ADC is determined by (2.8) for quantization noise only. However, in a well-designed system, the quantization noise should be lower than the receiver noise to optimize the receive sensitivity. Therefore, the effective number of bits is smaller than the available number of bits. A rule of thumb is to reduce the number of bits N by 2.5 [16] prior to applying (2.8) to calculate the ADC's dynamic range. For example, the dynamic range for a 10-

bit ADC is 47 dB corresponding to an effective number of bits of 7.5, and the quantization noise is ~16 dB below the receiver noise.

The overall system dynamic range will be the smallest one between that of the ADC and analog section of the receiver.

2.5 Radar Range Equation

The radar range equation or radar equation predicts the maximum range at which the radar can detect a specific target, and it is often used by radar engineers to design and analyze the radar system performance. Radar range equation is the counterpart of link budget analysis in communications.

If the radar transmits a waveform at power P_t (note: P_t is the peak power averaged over a cycle of the signal; referring to the detailed discussions in Chapter 3) through an isotropic antenna (uniform gain in all directions), the power density at distance R from the radar is equal to:

$$I_i = P_t / \left(4\pi R^2\right) \tag{2.21}$$

where $4\pi R^2$ is the surface area of an imaginary sphere of radius R. In reality, the antenna is never isotropic but directive. The power density at a direction where the transmit antenna has power gain G_t is:

$$I_D = P_t G_t / \left(4\pi R^2\right) \tag{2.22}$$

The target intercepts a fraction of the incident power and reradiates it to various directions. The power density reflected back to the radar (denoted by I_r) is determined by the radar cross section (RCS σ) of the target, which is defined by the following equation:

$$\sigma = 4\pi R^2 I_r / I_D \tag{2.23}$$

A more rigorous definition of RCS [17] is for far field:

$$\sigma = \lim_{r \to \infty} 4\pi r^2 \frac{I_r}{I_D} \tag{2.24}$$

The RCS has unit of area; however, it describes the target's ability to reflect power back to radar rather than its physical size. In fact the RCS is determined more by the target's shape, material, and aspect angle than by its physical size.

A portion of the returned power is received by the radar's receive antenna, and this received signal power is the product of the effective antenna aperture A_e and I_r:

$$P_r = I_r A_e = \frac{P_t G_t \sigma}{(4\pi)^2 R^4} A_e \tag{2.25}$$

The effective aperture can be obtained from (2.12), and (2.25) can be written as:

$$P_r = \frac{P_t G_t G_r \sigma \lambda^2}{(4\pi)^3 R^4} \tag{2.26}$$

where G_r is the power gain of the receive antenna if it is different from transmit antenna.

If there was no noise, radar could sense targets from infinite range. In reality, radar needs to detect targets among the competing noise, and the sources of this noise are from both the external environment and within the receiver, as discussed in Section 2.3. As a common practice, radar engineers take the receive antenna terminal as the reference point and consider all referred noise components, as shown in Figure 2.5. This is also the place where the receive power P_r is defined and measured. The noises generated at various stages of the receive system are all referred to this reference point for the ease of analysis. For example, the receiver noise temperature determined by (2.20) becomes $T_e L_r$ when referred back to the reference point to compensate for the loss of the transmission line. For a superheterodyne receive system, there are three main noise components [18], as shown in Figure 2.5: antenna noise temperature, transmission line noise temperature, and receiver noise temperature. The antenna output noise temperature is [19]:

$$T_a = \frac{0.876 T_a' - 254}{L_a} + T_0 \tag{2.27}$$

where T_a' is the noise temperature from solar and galaxy, which is about 300K for carrier frequency of 60 GHz.

The noise temperature of the transmission line is [20]:

$$T_r = T_{tr}\left(L_r - 1\right) \tag{2.28}$$

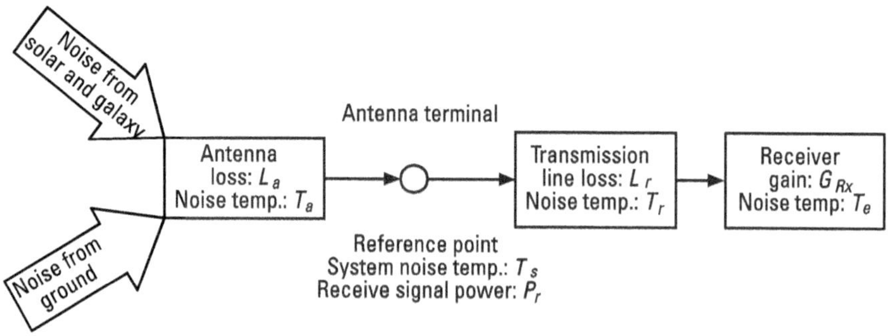

Figure 2.5 Sources and components of system noise temperature. L_a is the dissipative loss within the antenna, T_a is the antenna output noise temperature, T_s is the system noise temperature, L_r and T_r are, respectively, the loss and noise temperature of the transmission line.

where T_{tr} is the physical temperature of the transmission line and its value is usually set to be 290K. Equation (2.28) can be loosely understood in the following way. If the noise power at the input of the transmission line is $kT_{tr}B$, after passing through the component it will be reduced to $kT_{tr}B/L_r$. However, at the output of the transmission line the noise power cannot be lower than the thermal noise $kT_{tr}B$; this means the component itself generates certain noise power which at the output is equal to:

$$\Delta N = kT_{tr}B - \frac{kT_{tr}B}{L_r} = kT_{tr}B\left(1 - \frac{1}{L_r}\right) \tag{2.29}$$

When referring back to the input, the noise power becomes $\Delta N \bullet L_r$. According to the definition of (2.14), the noise temperature of the transmission line is then:

$$T_r = \Delta N \bullet L_r / kB = T_{tr}\left(L_r - 1\right) \tag{2.30}$$

At the reference point, the receiver noise temperature should be adjusted by the transmission line loss and the system noise temperature is:

$$T_s = T_a + T_r + T_e L_r \tag{2.31}$$

With (2.31), we can treat the whole receive system as an input resistor with noise temperature T_s and followed by an ideal receiver (no extra noise) having gain and loss characteristics of the actual system. For many civil applications, the radar needs to listen while transmitting to cover the very close range, and under such circumstance the transmitter's broadband noise also needs to be considered. The contribution due to this broadband noise can be estimated as:

$$T_t = \frac{P_t}{N_{tb} A_{tr} kB} \tag{2.32}$$

where N_{tb} is the transmitter power attenuation in the receive frequency band and A_{tr} is the transmit to receive channel isolation. For a well-designed system, N_{tb} >140 dBc and N_{tr}> 25 dB and the resultant transmitter wideband noise is negligible.

For homodyne systems such as LFM, (2.31) needs to also include the phase noise contributions. Phase noise is due to the phase fluctuations associated with the oscillator in waveform generator. It is different from thermal (white) noise in that phase noise decreases as the frequency increases from carrier, while thermal noise is independent of frequencies. Phase noise often includes flicker noise (also known as 1/f noise) for frequencies very close to the carrier. Referring to Figure 2.4, the phase noise in the received signal is correlated with that in the mixing signal in that they are both related to the same source of transmitted signal. The level of correlation is dependent on the time delay between the two signals, and this is known as range correlation effect. This effect significantly reduces the impact of phase noise on close in target detection in the received signals. According to [21], the baseband phase noise spectral density after mixing is:

$$S_{\Delta\phi}(f) = 2S_{\phi}(f)\left(1 - \cos(2\pi f \tau)\right) = S_{\phi}(f)\alpha \tag{2.33}$$

where $S_{\phi}(f)$ is the phase noise spectral density of the transmit (RF) signal, f is the frequency offset from the carrier, τ is the signal delay of the leakage signal or large reflection signal from enclosure of the radar, and α is the correlation attenuation. For example, if there is a strong leakage at 2 cm from the radar and the detection range of interests are up to 1 MHz (the target range of LFM radar is corresponding to beat frequency, as discussed in Chapter 3), the phase noise in the received signal is –61.5 dB below the transmit phase noise power according to (2.33). The phase noise temperature is given:

$$T_p = \frac{\alpha P_t}{A_p A_{tr} kB} \tag{2.34}$$

where A_p is the power ratio of phase noise to the carrier signal. Note A_p and α are both defined at given frequency offsets (corresponding to target range in LFM radar) from the carrier. Therefore, it may require iterative steps to find the appropriate SNR for maximum range analysis. Equation (2.31) should change to the following for LFM radar:

$$T_s = T_a + T_r + T_e L_r + T_p L_r \tag{2.35}$$

In order for the target to be detected among noise, a certain level of SNR is required. The minimum required SNR that can meet the desired probability of detection and false alarm rate is referred to as detectability factor D_x. The value of D_x is determined for various target models in [22]. Based on the earlier discussions, the SNR of the system can be expressed as:

$$SNR = \frac{P_r}{kT_s B} = \frac{P_t G_t G_r \sigma \lambda^2}{(4\pi)^3 R^4 k T_s B} \tag{2.36}$$

The SNR in (2.36) is the same as that of the output of a matched filter, which maximizes the detection performance. The maximum target detection range corresponding to a specific D_x is then:

$$R_{max}^4 = \frac{P_t G_t G_r \sigma \lambda^2}{(4\pi)^3 D_x k T_s B} \tag{2.37}$$

Equation (2.37) is the basic form of radar range equation. For many radar applications the product of bandwidth and waveform length is about one ($B\tau \approx 1$), and the radar equation can be formulated as:

$$R_{max}^4 = \frac{P_t \tau G_t G_r \sigma \lambda^2}{(4\pi)^3 D_x k T_s} = \frac{E_t G_t G_r \sigma \lambda^2}{(4\pi)^3 D_x k T_s} \tag{2.38}$$

where E_t is the transmit signal energy. The significance of (2.38) is that it is the radar transmits energy, which determines the maximum detection range.

There are many other factors that may need to be included in the radar range equation to have more precise prediction of the maximum range. The modified equation is:

$$R_{max}^4 = \frac{E_t G_t G_r \sigma \lambda^2 I_c I_n F_t^2 F_r^2}{(4\pi)^3 D_x k T_s L_t L_\alpha L_s} \tag{2.39}$$

where

I_c: Coherent integration gain such as Doppler processing where the signals are integrated with coherent phase. It is equal to the number of waveforms (pulses) integrated.

I_n: noncoherent integration gain. The signals are added in amplitude. The gain is a function of probability of detection and false alarm rate [22].

F_t^2, F_r^2: The propagation factors to account for the surface reflection and diffraction effects for transmit and receive paths. These values are often omitted in civil applications.

L_t: Transmission line loss between the transmitter output (where P_t is defined) and the transmit antenna terminal (where G_t is defined).

L_α: Atmospheric and precipitation attenuation. This value can be neglected for motion and gesture sensing radar. However, it is an important factor for automotive radar, and the two way loss is $L_\alpha = 10^{0.1k_\alpha R}$ with k_α being the attenuation coefficients in dB/km and R being the range from radar to target in km. The value of k_α is dependent on the weather type and the detailed discussions can be found in [23].

L_s: Other system losses such as processing losses due to taper application and detection loss due to CFAR detector.

It is worth pointing out that when the radar performance is limited by external noise or clutter echoes rather than receiver noise, the radar range equation takes on a completely different form from the equations presented in this section.

2.6 Radar System Integration

For many civil applications, radar needs to be integrated behind covers. For automotive it is the plastic facia, and for consumer electronics it is likely the display glass. The antenna efficiency and beampattern can be significantly influenced by the structures in front of it. How to optimize the placement of the radar and the materials and thickness of these structures can be complex, and there may not be any theoretic method available. An example is shown in Figure 2.6, where the radar on a chip is integrated into a cell phone with multiple layers above it. In practice, the antennas in the package are optimized with all candidate structures through a high fidelity simulation tool such as finite element method (FEM)–based electromagnetic solver. The best performance design is selected for the prototype measurement, and this process may iterate a few times before maturing to the final product design. The optimization objective may be not only the power gain but also minimizing the nulls if a wide field view is desired. Another important point is to keep the structure uniform in front of all receive antennas to maintain consistent phase response and achieve accurate angle information.

Another consideration is the coexistence with other electronic modules, and this is particularly challenging for radar integration into phones, watches, and other consumer electronics. The interference from other modules (e.g.,

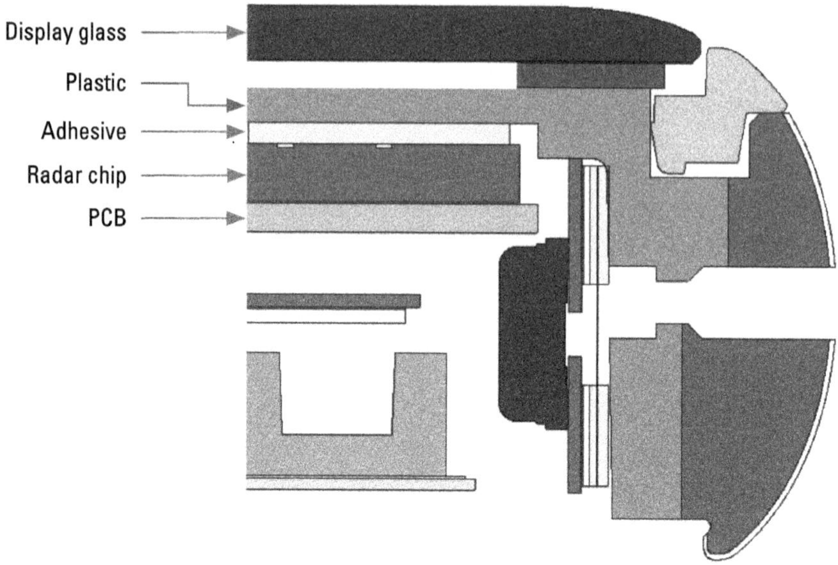

Figure 2.6 Radar integration into a cell phone.

WiFi, cellular, Bluetooth (BT)) can come through power supplies, and the well-designed power line filter and bypass capacitance can help minimize the interference effects. It has also been found that radar can be sensitive to supply voltage ripples, and a closely spaced LDO and bypass capacitance are helpful to mitigate the ripples.

The wireless charger can create strong interference through magnetic field interaction with the power supply of the radar chip, and it is recommended to keep these two separated from each other.

The radar chip can also be affected by the speaker module, whose vibration can lead to ghost targets. These ghost targets tend to show up in pairs with symmetry in Doppler domain, and this feature can be used to suppress these unwanted targets.

References

[1] Malevsky, S., and J. R. Long, "A Comparison of CMOS and BiCMOS mm-Wave Receiver Circuits for Applications at 60GHz and Beyond," in *Analog Circuit Design*, H. Casier, M. Steyaert, and A. H. M. van Roermund (eds.), Dordrecht, Netherlands: Springer, 2011.

[2] Zimmer, T., et al., "SiGe HBTs and BiCMOS Technology for Present and Future Millimeter-Wave Systems," *IEEE Journal of Microwaves*, Vol. 1, No. 1, 2021, pp. 288–298.

[3] Wang, H., and K. Sengupta, *RF and mm-Wave Power Generation in Silicon*, Academic Press, 2016.

[4] Maas, S. A., *Microwave Mixers*, Second Edition, Chapter 7, Norwood, MA: Artech House, 1993.

[5] Skolnik, M. I., *Introduction to Radar Systems*, Third Edition, Chapter 11.3, New York: McGraw-Hill, 2001.

[6] Ahmad, M. A., *High Speed Data Converters (Materials, Circuits and Devices)*, London: The Institution of Engineering and Technology, 2016.

[7] Kester, W., *Data Conversion Handbook*, Boston: Elsevier, 2005, pp. 677–691.

[8] Wang, J., et al., "Analysis of Concatenated Waveforms and Required STC," *IEEE Radar Conference*, 2008, pp. 1–6.

[9] Skolnik, M. I., *Radar Handbook*, Third Edition, Chapter 6.13, New York: McGraw-Hill, 2008.

[10] Smith, G. S., "A Direct Derivation of a Single-Antenna Reciprocity Relation for the Time Domain," *IEEE Transactions on Antennas and Propagation*, Vol. 52, No. 6, 2004, pp. 1568–1577.

[11] Kraus, J. D., *Antennas*, Third Edition, New York: McGraw-Hill, 2001.

[12] Saponara, S., et al., *Highly Integrated Low-Power Radars*, Norwood, MA: Artech House, 2014.

[13] Skolnik, M. I., *Introduction to Radar Systems*, Third Edition, Chapter 9.2, New York: McGraw-Hill, 2001.

[14] Johnson, J. B., "Thermal Agitation of Electricity in Conductors," *Physical Review Journals*, Vol. 32, 1928, pp. 97–109.

[15] Friis, H. T., "Noise Figures of Radio Receivers," *Proceedings of the IRE*, Vol. 32, No. 7, 1944, pp. 419–422.

[16] Stimson, G. W., et al., *Introduction to Airborne Radar*, Third Edition, Chapter 14.6, Edison, NJ: SciTech Publishing, 2014.

[17] Knott, E., et al., *Radar Cross Section*, Second Edition, Norwood, MA: Artech House, 1993.

[18] Blake, L. V., "A Guide to Basic Pulse-Radar Maximum-Range Calculation," *NRL Report 6930*, Naval Research Laboratory, 1969.

[19] Skolnik, M. I., *Radar Handbook*, Second Edition, Chapter 2.5, New York: McGraw-Hill, 1990.

[20] Blake, L. V., *Radar Range-Performance Analysis*, Norwood, MA: Artech House, 1986.

[21] Budge, M. C., and M. P. Burt, "Range Correlation Effects in Radars," *Record of the IEEE National Radar Conference*, 1993, pp. 212–216.

[22] Barton, D. K., *Radar System Analysis and Modeling*, Chapter 2, Norwood, MA: Artech House, 2005.

[23] Barton, D. K., *Radar System Analysis and Modeling*, Chapter 6, Norwood, MA: Artech House, 2005.

3

Radar Signal Model and Demodulation

Although the theory of radar may appear difficult and full of math derivations, the working principle is straightforward. A deterministic signal or waveform is transmitted through an antenna, and the returned signals are received and processed to obtain information about the environment around the radar. The received signals consist of returns from desired targets, echoes from unwanted targets, known as clutter, as well as interference and noise. How to effectively process the received signal to extract the correct information requires deep understanding of radar theory. In this chapter, we will walk you through the fundamentals of radar theory with simple language and mathematical derivations. The goal is not to target completeness of math, but to provide enough foundation for readers to build up their own understanding and enable them to solve real-world radar problems.

3.1 Signal Modeling

The general form of the transmitted signal can be modeled as the following:

$$s_{tx}(t) = u(t)cos(2\pi f_c t + \theta(t) + \theta_0) \tag{3.1}$$

where $u(t)$ is the signal envelope, which may contain amplitude modulation; f_c is the carrier frequency; $\theta(t)$ is the phase angle due to frequency/phase modulation; and θ_0 is an arbitrary initial phase. In some cases, θ_0 is considered a random value; however, more often θ_0 is treated as a constant, especially for coherent radar.

Very often radar engineers use the complex form of the signal model:

$$s_c(t) = g(t)e^{j(2\pi f_c t+\theta_0)} \tag{3.2}$$

which relates to $s_{tx}(t)$ with the following equation:

$$s_{tx}(t) = Re\left[s_c(t)\right] = \frac{1}{2}\left[g(t)e^{j(2\pi f_c t+\theta_0)} + g^*(t)e^{-j(2\pi f_c t+\theta_0)}\right] \tag{3.3}$$

where $g(t)$ is called the complex envelope of $s_{tx}(t)$; Re is a real part of the signal and * represents complex conjugate; j is the basic imaginary unit. Note that the complex signal is an abstraction that does not exist in the real world. The reason to use complex form modeling is for simpler mathematical manipulations. It is much easier to derive the response of $s_c(t)$ through a network and then take the real part of the results, compared to applying $s_{tx}(t)$ to obtain the response. For this reason, we adopt complex signal modeling in this book whenever appropriate.

Before proceeding to the next section, it is worth understanding how the complex signal model works and why it is easier to work with. $s_{tx}(t)$ can be expanded into the following:

$$s_{tx}(t) = u(t)cos(\theta(t))cos(2\pi f_c t+\theta_0) - u(t)sin(\theta(t))sin(2\pi f_c t+\theta_0) \tag{3.4}$$

Combining (3.2) to (3.4), we can obtain:

$$g(t) = u(t)cos(\theta(t)) + j \cdot u(t)sin(\theta(t)) = u(t)e^{j\theta(t)} \tag{3.5}$$

For most radar systems, the carrier frequency f_c is much greater than the bandwidth of amplitude envelope $u(t)$ and phase modulation $\theta(t)$; these systems are considered narrowband systems. In such systems, $g(t)$ consists of components varying much slower than $s_{tx}(t)$ due to carrier; for this reason, $g(t)$ is termed the complex envelope of $s_{tx}(t)$. The radar systems we will study later in this book can be treated as narrowband radar.

Substituting (3.5) to (3.2), we have:

$$s_c(t) = u(t)e^{j\theta(t)}e^{j(2\pi f_c t+\theta_0)} = u(t)e^{j(2\pi f_c t+\theta(t)+\theta_0)} \tag{3.6}$$

It is clear that the complex model just replaces the cosine function in the real model with the exponential function. If we denote $S_{tx}(f)$ as the Fourier transform of $s_{tx}(t)$, $G(f)$ as the Fourier transform of $g(t)$, and $S_c(f)$ as the Fourier transform of $s_c(f)$, the following are true according to (3.2) and (3.3):

$$S_c(f) = G(f - f_c)e^{j\theta_0} \tag{3.7}$$

$$S_{tx}(f)=\frac{1}{2}\left[G(f-f_c)e^{j\theta_0}+G^*(-f-f_c)e^{-j\theta_0}\right] \tag{3.8}$$

Illustrative spectrums are shown in Figure 3.1 to visualize the relationship between $S_{tx}(f)$, $G(f)$, and $S_c(f)$. Under the narrowband assumption, the bandwidth of spectrum $G(f)$ is much smaller than carrier frequency f_c. We should point out that the amplitude/phase/frequency modulation adopted in real radar systems will most likely give $G(f)$ nonzero spectrum residue even for $|f| > f_c$. However, these residues are relatively insignificant and can be neglected. A stricter math treatment can be found in [1], where the analytical signal concept [2] is utilized. The real signal model spectrum $S_{tx}(f)$ is a scaled version of $G(f)$ shifted to either $+f_c$ or $-f_c$. The complex signal model spectrum $S_c(f)$ is a scaled version of $G(f)$ shifted to f_c. There is no information lost by working with the complex signal model due to the facts that the negative frequency content of $S_{tx}(f)$ has no effects on the positive side of its spectrum and vice versa, and the positive and negative sides of the spectrums are images of each other. When working with the complex signal model, the complex envelope $g(t)$ is separated

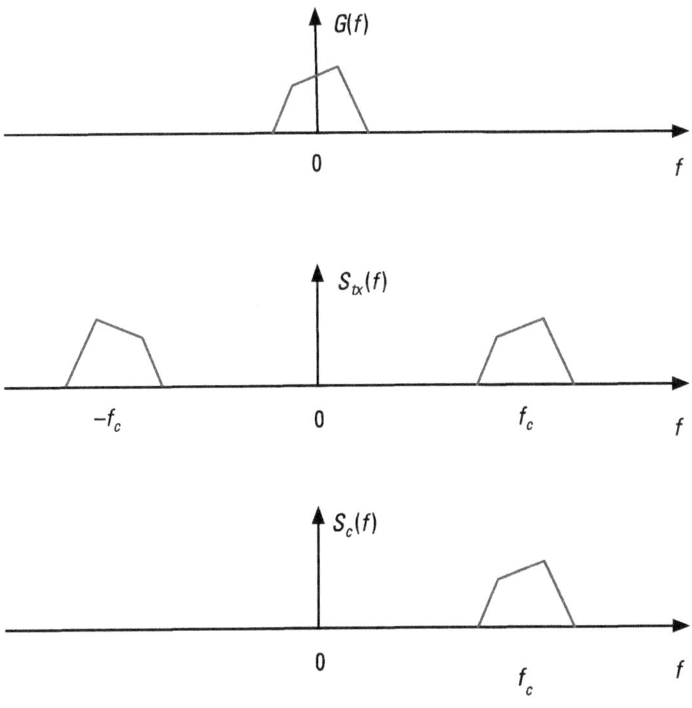

Figure 3.1 Magnitude spectrum of signal models.

from the carrier frequency term as shown in (3.2). This fact enables the analysis of an otherwise RF system response with a baseband system and baseband input signal $g(t)$ passing through it without considering the carrier. This significantly simplifies the simulation of radar systems with much lower sampling rate required (baseband versus RF). This will be clear in the following sections on receiver waveform demodulation.

In summary, it is much simpler to use complex signal modeling to analyze narrowband radar systems because we need to consider only the amplitude and phase modulation in a baseband system and can ignore the carrier frequency component.

The returned/scattered signals from targets are delayed and scaled versions of the transmitted signal. In many applications, the phase/frequency modulation of the returned signal changes from that of the transmitted signal, while the narrowband characteristics are maintained. In the following sections, the returned signals from a target will be studied first as if the target were a point scatterer and then as distributed scatterers.

3.1.1 Point Target

A point target or point scatterer means that its size is small compared to the radar resolution cell in range and cross-range. The target's physical scattering features are not resolvable. The complex return signal model of a point target can be represented as:

$$s_r(t) = g_r(t - t_0)e^{j(2\pi f_c(t-t_0)+\theta_0+\phi_0)} = \sqrt{\sigma}u(t - t_0)e^{j(2\pi f_c(t-t_0)+\phi(t-t_0)+\theta_0+\phi_0)} \quad (3.9)$$

where $g_r(t)$ is the return signal complex envelope and defined as:

$$g_r(t) = \sqrt{\sigma}u(t)e^{j\phi(t)} \quad (3.10)$$

where σ is the radar cross section (RCS) of the point target, and $\phi(t)$ is the phase modulation due to the point target reflection and the original modulation $\theta(t)$. Note $\phi(t)$ may not have the same form as $\theta(t)$ when the target has radial velocity. The time t_0 for the radar waveform to complete the two-way travel from the radar to the target and back is determined by the target distance R_0:

$$t_0 = 2R_0 / c \quad (3.11)$$

where c is speed of light. When there is one or a few scatterers with simple geometry within the radar resolution cell, the RCS of the point target may be determined with a closed form formula. However, in most cases the RCS is a random variable even for a point target. We will discuss the statistical model of RCS in a later chapter.

3.1.2 Distributed Target

A distributed target refers to a complex scatterer having dimensions larger than the resolution cell and allowing individual scatterers to be resolved. The complex return signal model of a stationary distributed target can be represented as:

$$s_r(t) = \int_{(R,\theta,\phi)} g_r(t,R',\theta',\phi') e^{j\left[2\pi f_c\left(t-\frac{c}{2R'}\right)+\theta_0(R',\theta',\phi')+\phi_0(R',\theta',\phi')\right]} dR'\,d\theta'\,d\phi' \quad (3.12)$$

$$s_r(t) = \int_{(R,\theta,\phi)} \sqrt{\sigma(R',\theta',\phi')}\,u\left(t-\frac{c}{2R'}\right) e^{j\left[2\pi f_c\left(t-\frac{c}{2R'}\right)+\phi(t,R',\theta',\phi')+\theta_0(R',\theta',\phi')+\phi_0(R',\phi',\phi')\right]} dR'\,d\theta'\,d\phi' \quad (3.13)$$

where R, θ, and ϕ, respectively, define the range, azimuth, and elevation span of the distributed target. If the distributed target can be represented with a number of dominating scatterers, the return signal model can be written as the sum of a finite number of returns from individual scatterers:

$$s_r(t) = \sum_i \sqrt{\sigma_i}\,u_i\left(t-\frac{c}{2R_i}\right) e^{j\left[2\pi f_c\left(t-\frac{c}{2R_i}\right)+\phi_i\left(t-\frac{c}{2R_i}\right)+\theta_{0,i}+\phi_{0,i}\right]} \quad (3.14)$$

An example is shown in Figure 3.2 for a hand target.

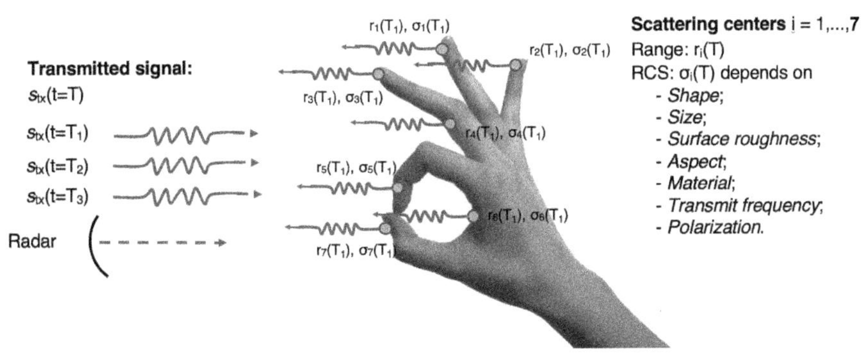

Figure 3.2 Discrete scattering center model of a hand.

3.2 Radar Waveforms and Demodulation

Depending on the specific amplitude modulation $u(t)$ and phase modulation $\theta(t)$, the radar transmit signal shown in (3.1) can have many different forms. At a high level, the transmitted signal can have continuous wave (CW) or pulsed waveforms. Within each of these categories, the transmitted signal can further have amplitude modulation, phase modulation, frequency modulation, or a combination of these. These modulations can be linear or nonlinear and random or deterministic. The pulsed waveform can have these modulations inside a single pulse or from pulse to pulse. Sometimes polarization can also be a modulation factor. Figures 3.3–3.5 illustrate a few common waveforms. If $\theta(t)$ is zero and $u(t)$ is the rectangular function, the waveform is a simple pulse with constant carrier frequency. If $u(t)$ is a constant, the corresponding waveform is a simple continuous waveform as shown in Figure 3.3. If $u(t)$ is constant and $\theta(t)$ varies in a way to result in the waveform's frequency linearly changing over time, the waveform is called linear frequency modulated (LFM). There are two forms of LFM, as shown in Figure 3.4: the LFM continuous wave and the LFM chirp train. There is another popular category of waveforms where the amplitude is constant and the phase of the waveform changes from time to time. A

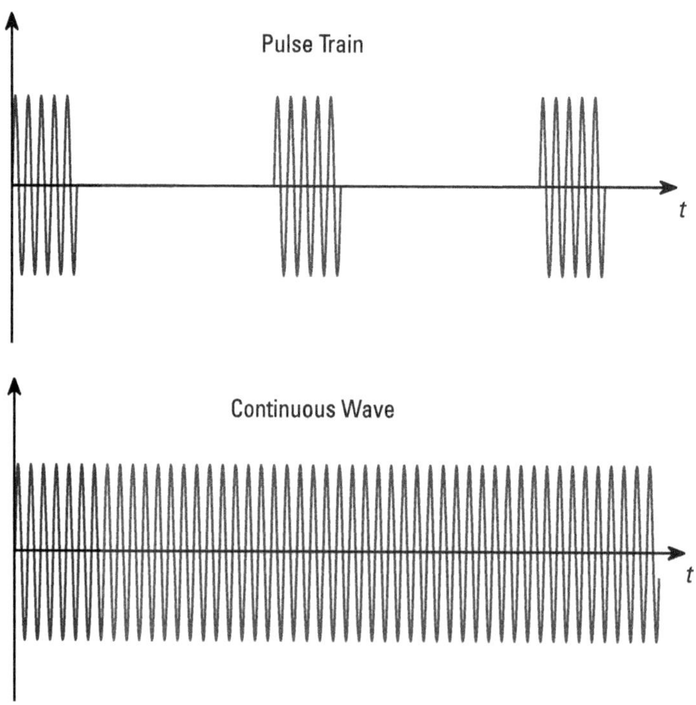

Figure 3.3 Simple pulse train and simple continuous waveform.

For simplicity, we consider the case where $n(t)$ is a white noise process with constant power spectral density over all frequencies [8]. If N_0 is the single-sided noise power spectral density specifying noise power per unit bandwidth and defined over the positive values of frequency, then the noise power spectral density at the output of the filter $h(t)$ is $|H(f)|^2 N_0/2$ [8]. The mean output noise power is simply the integral of its power spectral density across all frequencies:

$$N = \frac{N_0}{2}\int_{-\infty}^{+\infty} |H(f)|^2\, df \tag{3.17}$$

Note the reason for the ½ factor appearing before the integral in (3.17) is that the integration limits extend from $-\infty$ to $+\infty$ and N_0 is defined only for positive frequencies. Substituting (3.16) and (3.17) into (3.15), the objective function becomes:

$$R_{T_0} = \frac{\left|\int_{-\infty}^{+\infty} S_r(f) H(f) e^{j2\pi f T_0}\, df\right|^2}{\left(\frac{N_0}{2}\int_{-\infty}^{+\infty} |H(f)|^2\, df\right)} \tag{3.18}$$

The Cauchy–Schwarz inequality [9] states that if $f(x)$ and $g(x)$ are two square-integrable complex valued functions, then the following holds:

$$\left|\int f(x) g^*(x)\, dx\right|^2 \le \int |f(x)|^2\, dx \int |g(x)|^2\, dx = \int f(x) f^*(x)\, dx \int g(x) g^*(x)\, dx \tag{3.19}$$

and the equality is true if and only if $f(x) = kg(x)$ with k being a nonzero but arbitrary constant. If we let $f = H(f)$ and $g = S_r^*(f)e^{-j2\pi f T_0}$, and apply the Cauchy–Schwarz inequality to (3.18), then the following is true:

$$\begin{aligned} R_{T_0} &\le \int_{-\infty}^{+\infty} |H(f)|^2\, df \int_{-\infty}^{+\infty} |S_r(f)|^2\, df \Big/ \left(\frac{N_0}{2}\int_{-\infty}^{+\infty} |H(f)|^2\, df\right) \\ &= \int_{-\infty}^{+\infty} |S_r(f)|^2\, df \Big/ \left(\frac{N_0}{2}\right) \end{aligned} \tag{3.20}$$

Equation (3.20) shows that the ratio of instantaneous signal power to mean noise at the output of the filter is maximized when

$$H(f) = kS_r^*(f)e^{-j2\pi f T_0} \tag{3.21}$$

The equivalent time domain result is achieved by taking inverse Fourier transform of both sides of (3.21):

$$h(t) = ks_r^*(T_0 - t) \tag{3.22}$$

Since the factor k scales signal and noise at the same rate, it does not affect the final signal-to-noise ratio and can thus be simply set to 1. Equation (3.21) indicates that the magnitude of the matched filter's frequency response is the same as the amplitude spectrum of the input signal. This is the reason why the filter is named the matched filter. The phase of the matched filter frequency response equals the negative of the phase spectrum of the input signal with a linear phase shift. This observation helps explain how the matched filter works. In the frequency domain and at each frequency point, the multiplication of the input signal spectrum with its conjugate neutralizes its phase, leaving only the term $e^{-j2\pi f T_0}$ in the phase of the filter output signal. From the inverse Fourier transform, we know that this linear phase term effectively time delays all frequency components by the same time T_0. In the time domain, every frequency component of the output signal is perfectly aligned and in phase integrated at time T_0. This helps maximize signal power at T_0.

Equation (3.22) indicates that for any input signal, its corresponding matched filter is the input signal reversed in time, delayed by T_0, and complex conjugated. Figure 3.6 shows an example of an arbitrary input signal and its corresponding matched filter impulse response. For a matched filter to be realizable, it must be causal; that is, it cannot have any output values that depend on future input signal. This means $h(t) = 0$ for $t < 0$, and if the input signal starts at $t = 0$, it requires:

$$T_0 - \tau \geq 0 \tag{3.23}$$

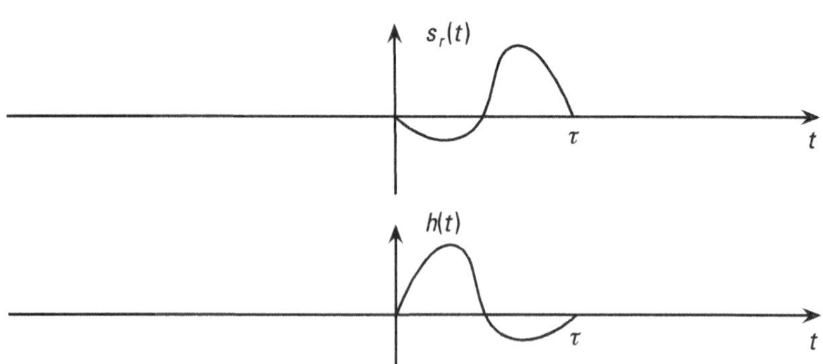

Figure 3.6 Received signal waveform and impulse response of the corresponding matched filter ($T_0 = \tau$).

where τ is the duration of the input signal. If the input signal has a start delay t_D, the realizable condition is:

$$T_0 - \tau - t_D \geq 0 \tag{3.24}$$

This case is illustrated in Figure 3.7, where the received signal starts at t_D with a duration of τ, and the peak ratio moment T_0 is chosen to be the same as $\tau + t_D$.

Since the matched filter is proportional to its input signal, the same matched filter also matches a scaled or time delayed version of the input signal. Scaling will not change the signal's shape, and it is obvious that the signal and all its nonzero scaled versions share the same matched filter. This conclusion is also true for time-delayed versions of the signal. This is important in practice since the time delay t_D is generally unknown. The matched filter can be designed for transmit signal or received signal $s_r(t)$ with zero delay, and it will automatically adapt to other received signals with arbitrary delays. Let us assume there is a second received signal $s_{r2}(t) = s_r(t - t_D)$ with $t_D > 0$ being an arbitrary delay. The spectrum of $s_{r2}(t)$ is:

$$S_{r2}(f) = S_r(f)e^{-j2\pi f t_D} \tag{3.25}$$

The corresponding matched filter of $s_{r2}(t)$ with maximum signal-to-noise ratio at T_2 is:

$$\begin{aligned} H_{r2}(f) &= S_{r2}^*(f)e^{-j2\pi f T_2} \\ &= S_r^*(f)e^{-j2\pi f T_0}e^{-j2\pi f(T_2 - t_D - T_0)} = H(f)e^{-j2\pi f(T_2 - (t_D + T_0))} \end{aligned} \tag{3.26}$$

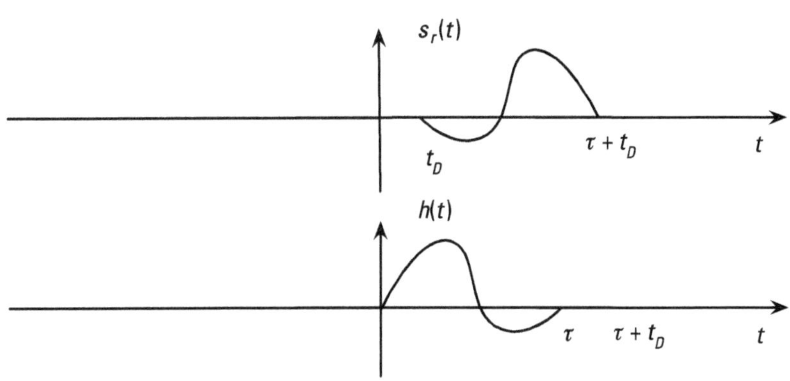

Figure 3.7 Received signal waveform and impulse response of the corresponding matched filter ($T_0 = \tau + t_D$).

Since T_2 can be set at any value as long as it meets the realizable condition, we can choose the peak signal-to-noise ratio moment $T_2 = t_D + T_0$, and the two matched filters are the same $H_{r2}(f) = H(f)$. In other words, the same matched filter can be applied to both received signals, and the peak signal-to-noise ratio will happen at time T_0 and T_2, respectively. This observation is of practical importance. It states that the matched filter can be designed for the transmit signal with $T_0 = \tau$ and applied to the received signals. If there are multiple returns in the received signals, there will be multiple peaks at the output of the matched filter. Each peak corresponds to its target return's time delay, which can be used to obtain the target distance with the formula $d = c(t - T_0)/2$, where c is the speed of light and t is the time corresponding to the peak.

According to Parseval's theorem [10], the energy of a signal/function in frequency domain equals to its energy in time domain, which gives:

$$\int_{-\infty}^{+\infty} \left|S_r(f)\right|^2 df = \int_{-\infty}^{+\infty} \left|s_r(t)\right|^2 dt = E \tag{3.27}$$

and E is the energy contained in signal $s_r(t)$. Combining (3.20) and (3.27):

$$R_{T_0} \le \frac{2E}{N_0} \tag{3.28}$$

When matched filter is applied, the maximum possible output peak signal to mean noise power ratio is $2E/N_0$. It states that the maximum signal-to-noise ratio only depends on the input signal's energy and the noise power spectral density N_0. The maximum ratio does not explicitly rely on the input signal's shape, duration, bandwidth, or modulation. For radar system analysis, we can assume that the radar receiver has a bandwidth roughly equal to the inverse of transmit signal duration:

$$B\tau \approx 1 \tag{3.29}$$

For applications where (3.29) is not true, a factor (either loss or gain) is applied to compensate for the delta. Combining (3.28) and (3.29), the following is obtained:

$$R_{T_0} \le \frac{2E}{N_0} = \frac{2P_r\tau}{N_0} = \frac{2P_r}{N} \tag{3.30}$$

where P_r is the received signal peak power. Equation (3.30) provides the foundation for radar range equation and link budget analysis as discussed in Chapter 2. It justifies the way that the signal-to-noise ratio are defined in radar range equation. It should be noted that the signal-to-noise ratio in (3.30) is twice of that in radar range equation. The reason is that the peak signal power in (3.30) is instantaneous peak signal power, which only occur at the peak of an RF or IF cycle. This is not possible in practice. The more appropriate peak signal power should be the power averaged over a cycle of the signal, which is half of what is in (3.30) for a rectangular sinewave pulse [11]. This is why the E/N_0 or P_r/N is ordinarily adopted as the signal-to-noise ratio in radar range equation by radar engineers. Another way to look at the problem is through phase shift. The matched filter can perfectly match the transmit signal; however, in practice the received signal often has unpredicted phase shift compared to the transmitted signal. Therefore, the output of matched filter is often not going to have the expected peak instantaneous power. More realistic is to average the peak instantaneous power over 2π phase cycle.

The matched filter discussed so far is based on input white noise assumption. For nonwhite (colored) noise, the matched filter consists of a cascade of two filters. The first filter is a prewhitening filter, which makes the noise spectrum uniform. The second filter matches the signal at the output of the first filter. Detailed design and analysis can be found in [1, 12], however the colored noise matched filter is rarely applicable in radar in that noise is generally uniform over the bandwidth of a radar receiver.

3.2.2 Ambiguity Function

Until now the matched filter is discussed under the assumption that the received signal is a time-delayed replica of the transmitted signal reflected from a stationary target. This assumes there is no Doppler shift and the received signal has the same frequency as the transmitted signal. Doppler is a shift in frequency that is introduced by relative radial velocity between the radar and targets. The target is moving in many radar applications, and the received signal has a Doppler frequency shift. Under these circumstances, the received signal is no longer matched by the matched filter.

It is important to study the matched filter output as a function of both time delay and Doppler frequency, which is key to understand and evaluate the properties of a radar waveform such as resolution, accuracy, sidelobes, and ambiguities in range and radial velocity. The tool to achieve this goal is ambiguity function originated by Woodward [13]. Ambiguity function is critical to design and analyze radar waveforms. Before we introduce ambiguity function, it is worth understanding what Doppler effect is and how this effect modulates the received signals.

The Doppler effect is a shift in the frequency of a radiated wave that is reflected or received by a target having relative radial motion. Note the tangential motion between radar and target will not result in Doppler effect. The relativistic Doppler effect [14, 15] states that if the radar transmitter and receiver are colocated and a target is approaching with a radial velocity of v, the received frequency is:

$$f_r = f_c \frac{1+v/c}{1-v/c} \tag{3.31}$$

where f_c is the transmit signal frequency and c is speed of light. The strict derivation of Doppler effect requires special theory of relativity, which is beyond the focus of this book. We will explain Doppler effect from an intuitive and simplified point of view, and the result is just as accurate for general radar applications. Let us assume the radar receiver and transmitter are stationary and colocated, a single tone with frequency f_c is transmitted, and there is a point target approaching radar at a constant radial velocity of v, as shown in Figure 3.8. The slightly curved vertical lines represent wavefronts of plane waves. Every point on the wavefronts has the same phase of the wave field. The two curved lines are used to represent two continuous wave crests. The wave crests are wavefronts with maximum field intensity in the positive direction.

The distance between these two wave crests is one wavelength and time difference is one period of $T_c = 1/f_c$. The goal is to find the received signal's period $T_r = T_{r2} - T_{r1}$, where T_{r1} is the time for the first wave crest to travel back to radar receiver after reflecting from target, and T_{r2} is time for the second wave crest to come back. The starting time of the first wave crest is 0 and the second wave crest is T_c. The two wave crests travel different distance due to the motion of the target and the distance delta can be determined:

$$\Delta d = cT_{r1} - c\left(T_{r2} - T_c\right) \tag{3.32}$$

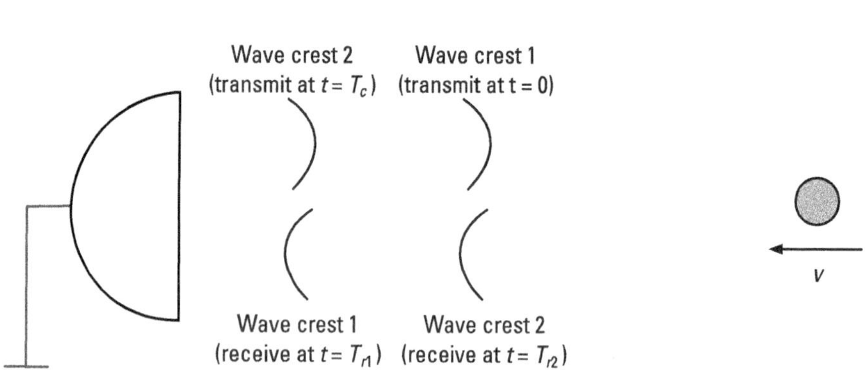

Figure 3.8 Two wave crests on the transmitted wave that are separated by one carrier cycle in time are reflected back to the receiver by constant moving point target.

The distance delta Δd should also equal to twice of the distance the target has moved, from the moment of the first wave crest reaching the target to the moment of second wave crest reaching the target:

$$\Delta d = 2\left[v\left(\left(T_{r2} - T_c\right)/2 + T_c\right) - \left(vT_{r1}/2\right)\right] \tag{3.33}$$

where $(vT_{r1}/2)$ is how far the target has moved from the beginning to when it meets the first wave crest, $v((T_{r2} - T_c)/2 + T_c)$ is how far the target has moved from the beginning to when it meets the second wave crest, and the factor 2 is to account for two way travel. Combining (3.32) and (3.33), and considering $T_r = T_{r2} - T_{r1}$ the following can be obtained:

$$T_r = \frac{c - v}{c + v} T_c \tag{3.34}$$

or

$$f_r = \frac{c + v}{c - v} f_c = f_c \frac{1 + v/c}{1 - v/c} \tag{3.35}$$

It can be seen that (3.35) is identical to (3.31). The change in received frequency relative to the transmitted frequency is called the Doppler frequency:

$$f_d = f_r - f_c = \frac{2v/c}{1 - v/c} f_c \tag{3.36}$$

When $v \ll c$, the Doppler frequency can be approximated:

$$f_d \approx \frac{2v}{c} f_c = \frac{2v}{\lambda} \tag{3.37}$$

where λ is the wavelength of the transmitted signal. The Doppler frequency will have a positive sign for approaching targets as shown in the example of Figure 3.8, and negative sign for moving away targets. Only relative radial velocity will result in Doppler effects, and the tangential velocity can be neglected. There are various examples for Doppler frequency calculation in [16] for different motion patterns of both radar and targets.

The transmit signal generally has a finite bandwidth B_t, which also contributes to the Doppler frequency on top of the carrier frequency. According to (3.37) the received signal bandwidth is:

$$B_r = \frac{2v}{c} B_t + B_t = \frac{2v}{\lambda f_c} B_t + B_t \tag{3.38}$$

For most radar applications, the Doppler effects due to bandwidth can be neglected in that the bandwidth is much smaller than carrier frequency. Under such circumstances, the radar return signal model (3.9) can be modified into the following [17] for a point target with t_0 delay and Doppler frequency f_d:

$$\begin{aligned} s_r(t) &= \sqrt{\sigma} u(t - t_0) e^{j(2\pi(f_c + f_d)(t-t_0) + \theta(t-t_0) + \theta_0 + \phi_0)} \\ &= \sqrt{\sigma} g(t - t_0) e^{j(2\pi(f_c + f_d)(t-t_0) + \theta_0 + \phi_0)} \end{aligned} \tag{3.39}$$

and $g_r(t) = \sqrt{\sigma} g(t)$ because $\phi(t)$ in (3.10) equals to $\theta(t)$ in (3.6). The following is also true:

$$s_r(t) = \sqrt{\sigma} s_c(t - t_0) e^{j(2\pi f_d (t-t_0) + \phi_0)} \tag{3.40}$$

The matched filter designed for the transmit signal (3.6) is:

$$\begin{aligned} h(t) &= k s_c^*(T_0 - t) = k u(T_0 - t) e^{-j(2\pi f_c (T_0 - t) + \theta(T_0 - t) + \theta_0)} \\ &= k g^*(T_0 - t) e^{-j(2\pi f_c (T_0 - t) + \theta_0)} \end{aligned} \tag{3.41}$$

k is an arbitrary scaler and will be set as $1/\sqrt{\sigma}$ to simplify the analysis. According to (3.39) and (3.41), the filter (matched to transmit signal) output of the received signal is as follows with time delay t_0 and Doppler frequency f_d being variables:

$$y_r(t) = \int_{-\infty}^{+\infty} g^*(T_0 - \tau) e^{-j(2\pi f_c (T_0 - \tau) + \theta_0)} g(t - \tau - t_0) e^{j(2\pi (f_c + f_d)(t - \tau - t_0) + \theta_0 + \phi_0)} d\tau \tag{3.42}$$

We are interested in the filter output at the nominal moment $t = T_0$ for various time delay t_0 and Doppler frequency f_d:

$$\begin{aligned} & y_r(T_0; t_0, f_d) = \\ & \int_{-\infty}^{+\infty} g^*(T_0 - \tau) e^{-j(2\pi f_c (T_0 - \tau) + \theta_0)} g(T_0 - \tau - t_0) e^{j(2\pi (f_c + f_d)(T_0 - \tau - t_0) + \theta_0 + \phi_0)} d\tau \end{aligned} \tag{3.43}$$

If changing integral variable τ to $\tau' = T_0 - \tau - t_0$, (3.43) becomes:

$$y_r(T_0;t_0,f_d) = -e^{-j(2\pi f_c t_0 - \phi_0)} \int_{-\infty}^{+\infty} g(\tau') g^*(\tau' + t_0) e^{j2\pi f_d \tau'} d\tau' \tag{3.44}$$

The magnitude of $y_r(T_0;t_0, f_d)$ is determined by the integral term, whose squared magnitude is defined as ambiguity function:

$$|\chi(t_0,f_d)|^2 = \left| \int_{-\infty}^{+\infty} g(\tau') g^*(\tau' + t_0) e^{j2\pi f_d \tau'} d\tau' \right|^2 \tag{3.45}$$

An alternative form of ambiguity function is:

$$|\chi(t_0,f_d)|^2 = \left| \int_{-\infty}^{+\infty} s_c(\tau') s_c^*(\tau' + t_0) e^{j2\pi f_d \tau'} d\tau' \right|^2 \tag{3.46}$$

where positive t_0 indicates a target is further away than the expected target, and positive f_d means an approaching target. Readers may be confused with why it is $s_c^*(\tau' + t_0)$ instead of $s_c^*(\tau' - t_0)$. The part $s_c(\tau')e^{j2\pi f_d \tau'}$ in (3.46) can be considered as a return signal with Doppler shift f_d and a time delay t_0 relative to the matched filter. The matched filter response should be shifted to the left of the time axis by t_0 (thereafter taking the form of $s_c^*(\tau' + t_0)$) to make the relative time delay of $s_c(\tau')e^{j2\pi f_d \tau'}$ to be t_0.

If $t_0 = 0$ and $f_d = 0$, the matched filter of the transmitted signal also matches the received signal and (3.45) becomes:

$$|\chi(0,0)|^2 = \iint \left| \int_{-\infty}^{+\infty} g(\tau') g^*(\tau') d\tau' \right|^2 = E^2 \tag{3.47}$$

According to Cauchy–Schwarz inequality (3.19), it is straightforward to show:

$$|\chi(t_0,f_d)|^2 \le |\chi(0,0)|^2 = E^2 \tag{3.48}$$

The volume of the ambiguity function is a constant and determined by:

$$volume = \iint_{-\infty}^{+\infty} |\chi(t_0,f_d)|^2 \, dt_0 df_d = E^2 \tag{3.49}$$

This property is referred to as volume invariance and it can be proved as follows:

$$volume = \iint_{-\infty}^{+\infty} dt_0 df_d \int_{-\infty}^{+\infty} g(\tau')g^*(\tau'+t_0)e^{j2\pi f_d \tau'}d\tau' \int_{-\infty}^{+\infty} g^*(\tau)g(\tau+t_0)e^{-j2\pi f_d \tau}d\tau \quad (3.50)$$

After reshuffling the order of integral, the following is true:

$$volume = \iint_{-\infty}^{+\infty} \int_{-\infty}^{+\infty}\int_{-\infty}^{+\infty} e^{j2\pi f_d(\tau'-\tau)} df_d g(\tau')g^*(\tau'+t_0)g^*(\tau)g(\tau+t_0)d\tau' d\tau dt_0 \quad (3.51)$$

According to Fourier transforms property, $\int_{-\infty}^{+\infty} e^{j2\pi f_d(\tau'-\tau)}df_d = \delta(\tau'-\tau)$ and $\delta(\tau' - \tau)$ is Dirac delta function with only nonzero values at $\tau' = \tau$. (3.51) can be simplified into:

$$volume = \iint_{-\infty}^{+\infty} \int_{-\infty}^{+\infty} \delta(\tau'-\tau)g^*(\tau'+t_0)g(\tau')d\tau' g^*(\tau)g(\tau+t_0)d\tau dt_0 \quad (3.52)$$

Realizing $\int_{-\infty}^{+\infty} \delta(\tau'-\tau)g^*(\tau'+t_0)g(\tau')d\tau' = g^*(\tau+t_0)g(\tau)$, then (3.52) becomes:

$$volume = \iint_{-\infty}^{+\infty} g(\tau+t_0)g^*(\tau+t_0)g(\tau)g^*(\tau)d\tau dt_0 \quad (3.53)$$

Noticing $\int_{-\infty}^{+\infty} g(\tau+t_0)g^*(\tau+t_0)dt_0 = E$ for any value of τ, (3.53) becomes:

$$volume = E\int_{-\infty}^{+\infty} g(\tau)g^*(\tau)d\tau = E^2 \quad (3.54)$$

The third important property of ambiguity function is its symmetry relation as:

$$\left|\chi(t_0, f_d)\right|^2 = \left|\chi(-t_0, -f_d)\right|^2 \quad (3.55)$$

This property can be readily proven from the definition of ambiguity function:

$$\left|\chi(-t_0, -f_d)\right|^2 = \left|\int_{-\infty}^{+\infty} g(\tau')g^*(\tau'-t_0)e^{-j2\pi f_d \tau'}d\tau'\right|^2 \quad (3.56)$$

Substituting the integral variable with $\tau = \tau' - t_0$, the following is true:

$$\left|\chi(-t_0, -f_d)\right|^2 = \left|\left(e^{j2\pi f_d t_0}\int_{-\infty}^{+\infty} g(\tau)g^*(\tau+t_0)e^{j2\pi f_d \tau}d\tau\right)^*\right|^2 = \left|\chi(t_0, f_d)\right|^2 \quad (3.57)$$

The volume invariance property states that the total volume of an ambiguity function is bounded by its corresponding waveform energy and independent of the waveform modulation. This restriction is well explained in [18] by imagining there is a fixed amount of sand in a box. If a narrow pile of sand is kept at the origin for good resolution and accuracy, the remaining sand must build up somewhere else (resulting in ambiguity or sidelobes in range and/or Doppler) due to the limit (E^2) on the height of the sand. On the other hand, the peaks outside of the origin can be controlled through either increasing the pile size at the origin or placing the peaks in a region that is less important, such as Doppler space beyond interests. When designing radar waveforms, we must consider the tradeoffs among resolution, accuracy, and ambiguities.

Due to these ambiguity function properties discussed so far, we cannot arbitrarily choose a function as the desired ambiguity function in that there may not be a valid waveform associated with it. For example, an ideal ambiguity function would be an impulse function as shown in Figure 3.9. The infinitesimal thickness at the origin allows the Doppler and range to be determined simultaneously with high fidelity. The associated waveform will then provide the best resolution and accuracy with no ambiguity, except that it cannot be practically obtained. The reason is obvious that this ideal ambiguity function does not meet the volume invariance and maximum bound constraints.

Before proceeding to the specific waveforms suitable for motion and gesture sensing, it is worthwhile showing the ambiguity functions of a simple pulse to illustrate the points we have discussed so far.

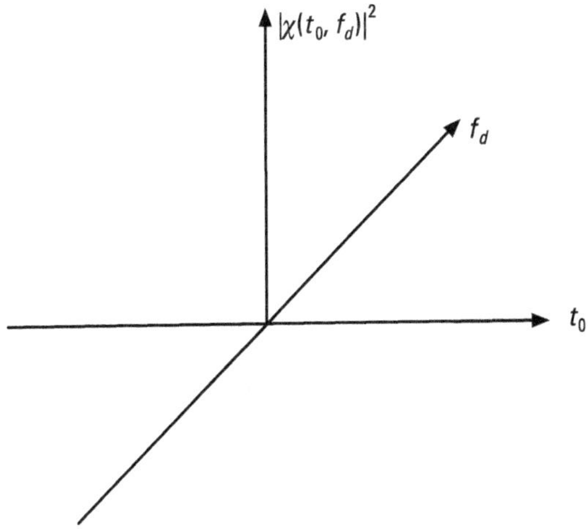

Figure 3.9 Ideal ambiguity function.

Let us first consider a simple normalized (energy $E = 1$) rectangular pulse centered at the origin as shown in Figure 3.10 and described by the following equation:

$$g(t) = \begin{cases} 1/\sqrt{\tau} & \dfrac{-\tau}{2} \le t \le \dfrac{\tau}{2} \\ 0, & elsewhere \end{cases} \tag{3.58}$$

According to (3.45), the corresponding ambiguity function of (3.58) is:

$$|\chi(t_0, f_d)|^2 = \begin{cases} \left|sin\left[\pi f_d\left(\tau - |t_0|\right)\right] / \tau\pi f_d\right|^2, & -\tau \le t_0 \le \tau \\ 0, & elsewhere \end{cases} \tag{3.59}$$

When there is no Doppler shift ($f_d = 0$), the ambiguity function is the squared matched filter output by utilizing L'Hospital's rule [19]:

$$|\chi(t_0, 0)|^2 = \begin{cases} \left(\tau - |t_0|\right)^2 / \tau^2, & -\tau \le t_0 \le \tau \\ 0, & elsewhere \end{cases} \tag{3.60}$$

Where there is no time delay ($t_0 = 0$), the ambiguity function becomes squared Sinc function:

$$|\chi(0, f_d)|^2 = |sin(\pi f_d \tau)/(\pi f_d \tau)|^2 \tag{3.61}$$

The 3-D plot (referred to as the ambiguity diagram in [20]) of the ambiguity function is shown in Figure 3.11 with normalized axis. Figure 3.12 is the contour plot of the ambiguity diagram. Figure 3.13 illustrates the cut along time delay axis t_0 with $f_d = 0$, and Figure 3.14 shows the cut along Doppler axis f_d with $t_0 = 0$. It can be observed that the ambiguity function is bounded by the maximum value at the origin with $E = 1$. The symmetry property is clear in Figure 3.12. The narrower the mainlobe, the better the measurement accuracy

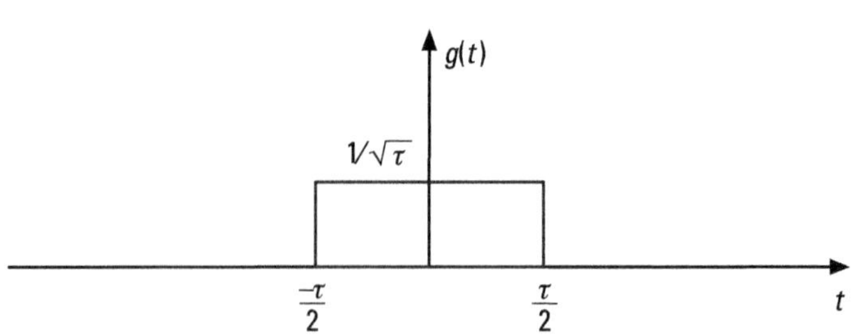

Figure 3.10 Waveform of a simple pulse.

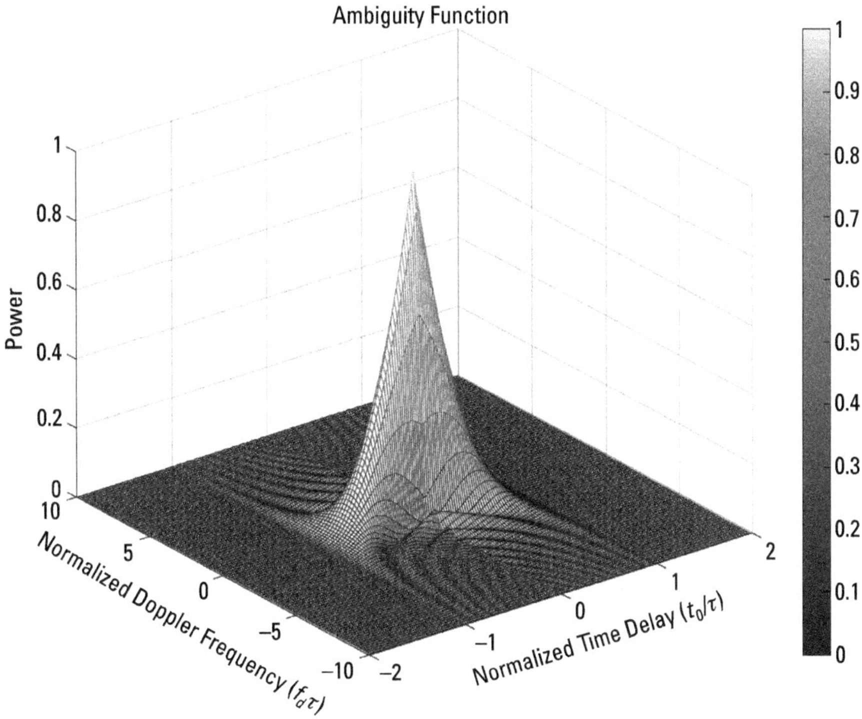

Figure 3.11 Ambiguity function diagram.

is going to be. From Figures 3.13 and 3.14, it is obvious that the time delay accuracy is proportional to τ and the Doppler frequency accuracy is proportional to $1/\tau$. This indicates that time delay and Doppler frequency are ambiguous for this waveform, and we cannot measure both time delay and Doppler in high accuracy with a simple unmodulated pulse.

Figure 3.15 illustrates the ambiguity function's contour plots of two simple pulses with different durations. The top figure shows the results of a relative long pulse with $\tau = 50\ \mu s$, and the bottom figure shows the results of a relative short pulse with $\tau = 5\ \mu s$. With the long pulse, the Doppler measurement is improved (narrower mainlobe along Doppler axis), however with poor time delay accuracy. On the contrary, with the short pulse we can have good time delay measurement (narrower mainlobe along time delay axis) but poor Doppler measurement.

If the short pulse in Figure 3.15 is reduced to nearly impulse, the waveform is ultrawideband. Its ambiguity function becomes a thin line along Doppler axis, which means the range measurement accuracy is great with no Doppler measurement capability. In order to measure both range and Doppler accurately, a train of very short pulses are required.

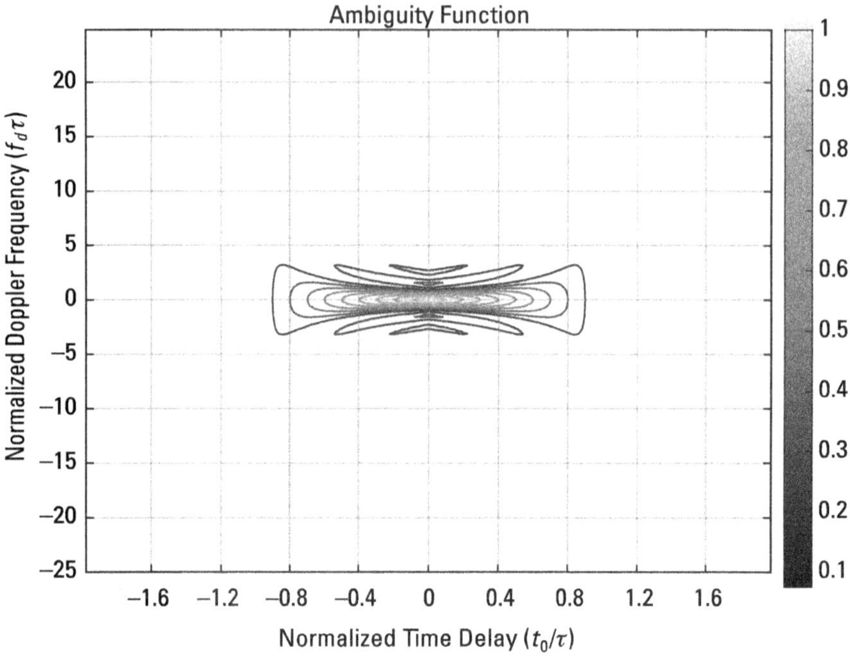

Figure 3.12 Ambiguity function contour.

Another important information from ambiguity function determines if the waveform is Doppler tolerant. The very short pulse shown in Figure 3.15 has bad Doppler accuracy; however, it is Doppler tolerant in that a single matched filter (matched for signals without Doppler shift) will produce strong output even with the input signal having significant Doppler shift. On the other hand, a single matched filter (matched for signals without Doppler shift) of long pulse will have greatly reduced output for these returns with significant Doppler shift. This means targets with high radial velocity are not likely to be detected. This is the reason that some conventional radars adopt a bank of matched filters (each matched to a different Doppler shift) to cover the expected range of Doppler frequencies.

If the duration of the long pulse in Figure 3.15 is extended to infinity or very large, it becomes unmodulated continuous waveform and its ambiguity function changes to a very thin line along time delay axis. This means the Doppler measurement accuracy is superior (nearly perfect) but with nearly no range measurement capability. This waveform has been widely adopted in radar guns for law enforcement, tennis and baseball speed measurement for competition sports, and vital signs measurement when only Doppler/velocity is of concerns.

In Section 3.1 we have explained that the carrier frequency can be ignored with the adoption of complex signal modeling without affecting the analysis

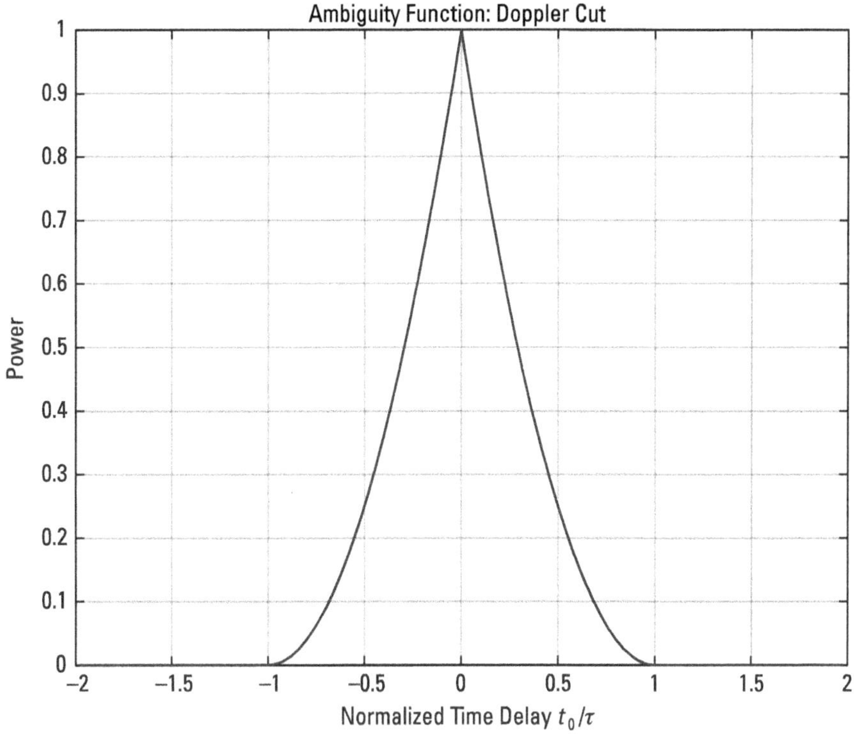

Figure 3.13 Cut along time delay with Doppler frequency $f_d = 0$.

and simulation results. This is why only the envelope of the signals is considered when we discuss these examples.

3.3 Frequency Modulated Waveforms

Simple waveform such as a rectangular pulse has the ambiguity tradeoff between time delay and Doppler shift. Time delay accuracy determines the range measurement accuracy. In practice, it is more common to employ waveforms with frequency or phase modulations. In this section we will discuss frequency modulated signals with particular focus on linear frequency modulation, and in the next section we will focus on phase coded waveforms. The LFM is the most popular frequency modulated signals in practice due to its simple circuits' requirements.

There are two forms of LFM waveforms as shown in Figure 3.4 in practical radar systems, and both can be found in civil applications. The first one is linear frequency modulated continuous waveforms (linear FMCW) as illustrated in Figure 3.16 for its time/frequency behavior. Linear FMCW has

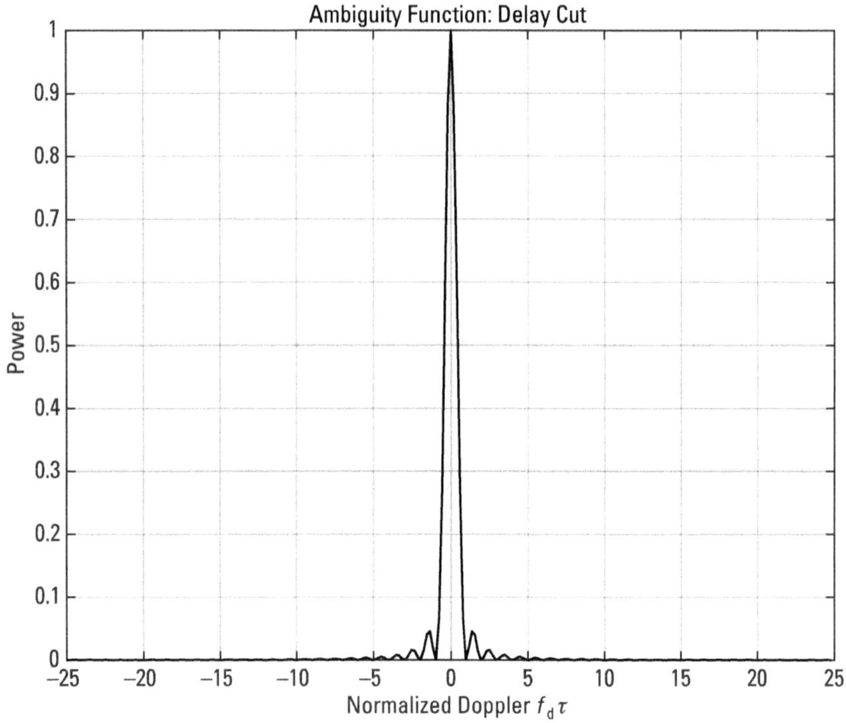

Figure 3.14 Cut along Doppler frequency with time delay $t_0 = 0$.

even simpler requirements on analog circuits with very low cost. It, however, requires complex algorithms to resolve Doppler/range ambiguity, which poses a challenge when there are a lot of targets in the field of view. This leads to the recent adoption of LFM chirp train waveforms, which can better handle multiple targets. This waveform is also being referred to as fast chirp, as illustrated in Figure 3.17 for its time/frequency behavior. In fact, we will demonstrate that LFM chirp train requires similar signal processing techniques as these in pulsed waveform system when certain conditions are met. The LFM chirp train is widely adopted by modern automotive radar systems as well as motion and gesture control radars.

Figure 3.16 shows an example up-chirp FMCW waveform, which means the chirp starts with lower carrier frequency f_1 and linearly ramps up to higher carrier frequency f_2 during the overall waveform duration τ. Alternatively, the carrier frequency can decrease over the duration of the chirp known as down-chirp. In Figure 3.17 there are a few chirps during the same duration τ, and each chirp sweeps the same bandwidth as that in Figure 3.16 but with much higher sweeping slope/rate. This is the reason it is sometimes referred to as fast chirp [21], and this waveform has been widely applied in modern automotive

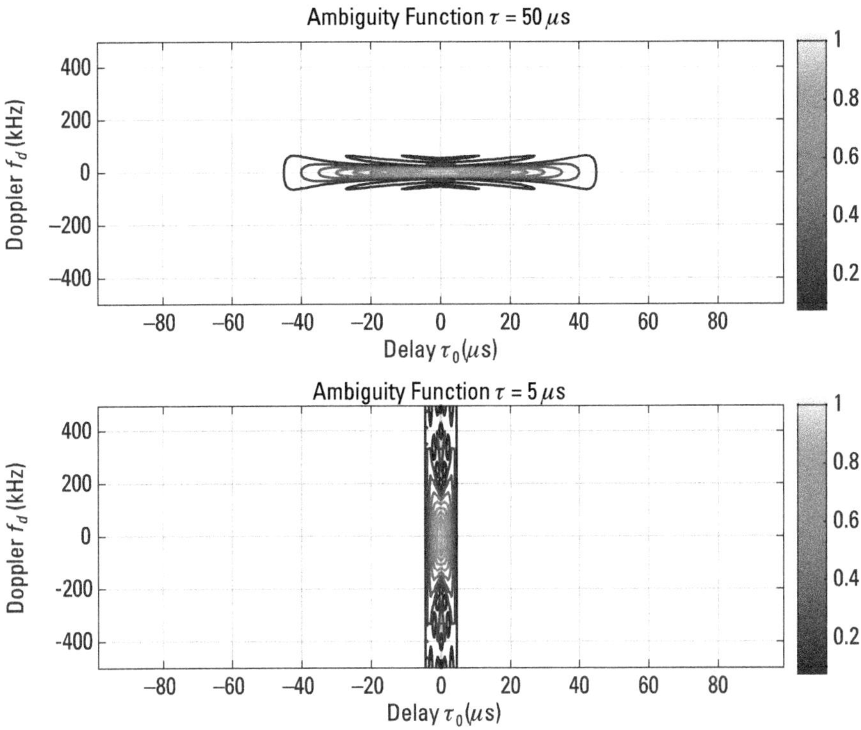

Figure 3.15 Ambiguity function long pulse versus short pulse.

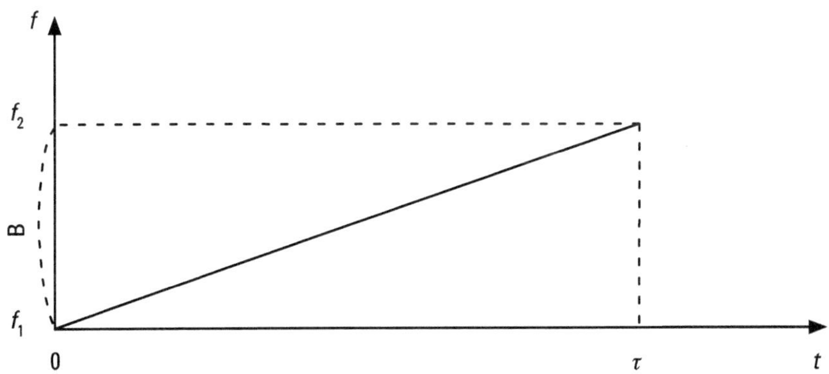

Figure 3.16 Example of FMCW waveforms in time frequency domain. B is the sweep bandwidth.

radars [22–24]. Note the gaps between chirps in Figure 3.17 are not mandatory. It can be technically zero duration or relatively large depending on the specific detection requirements as discussed in the following sections.

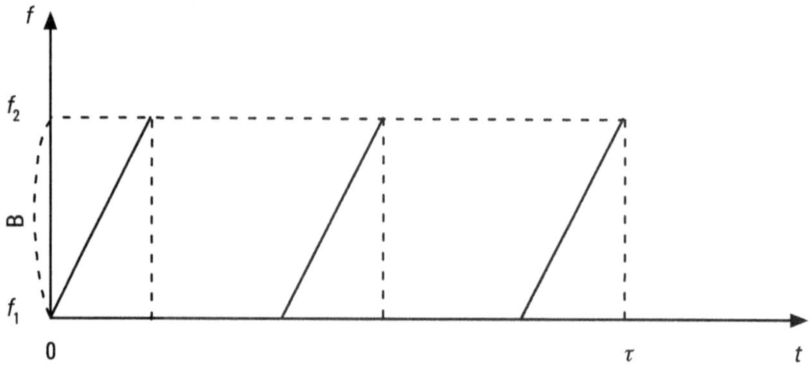

Figure 3.17 Example of LFM chirp chain waveforms in time frequency domain. B is the sweep bandwidth.

3.3.1 Conventional FMCW Waveforms

A LFM chirp can be expressed as:

$$s_{tx}(t) = A\cos\left(2\pi f_c t + 2\pi\beta t^2/2 + \theta_0\right) \quad -\tau/2 \le t \le \tau/2 \tag{3.62}$$

where A is the constant envelope, f_c is the carrier frequency, β is the sweep slope, and θ_0 is an arbitrary initial phase. The instantaneous frequency of the signal is the derivative of its phase against time:

$$f_i(t) = f_c + \beta t \quad -\tau/2 \le t \le \tau/2 \tag{3.63}$$

The sweep bandwidth B is related to the sweep slope β and sweep duration τ as follows

$$B = f_i(\tau/2) - f_i(-\tau/2) = \beta\tau \tag{3.64}$$

To simplify the analysis, let us define the normalized complex envelope (energy $E = 1$) of the LFM chirp as:

$$g(t) = 1/\sqrt{\tau}\, e^{j\pi\beta t^2} \tag{3.65}$$

According to (3.45), the corresponding ambiguity function of (3.65) is:

$$|\chi(t_0, f_d)|^2 = \begin{cases} \left|\sin\left[\pi(\beta t_0 + f_d)(\tau - |t_0|)\right]/\left[(\beta t_0 + f_d)\pi\tau\right]\right|^2, & -\tau \le t_0 \le \tau \\ 0, & \textit{elsewhere} \end{cases} \tag{3.66}$$

When there is no Doppler shift ($f_d = 0$), the ambiguity function of LFM is the squared matched filter output:

$$|\chi(t_0,0)|^2 = \begin{cases} \left|sin\left[\pi\beta t_0\left(\tau-|t_0|\right)\right]/(\pi\beta t_0\tau)\right|^2, & -\tau \le t_0 \le \tau \\ 0, & elsewhere \end{cases} \tag{3.67}$$

Where there is no time delay ($t_0 = 0$), the ambiguity function becomes squared Sinc function:

$$|\chi(0,f_d)|^2 = |sin(\pi f_d\tau)/(\pi f_d\tau)|^2 \tag{3.68}$$

The ambiguity diagram contour for an up-chirp ($\beta > 0$) is shown in Figure 3.18 with normalized axis. Figure 3.19 illustrates the cut along time delay axis t_0 with $f_d = 0$, and Figure 3.20 shows the cut along Doppler axis f_d with $t_0 = 0$. Figure 3.20 indicates that the Doppler frequency accuracy is proportional to $1/\tau$, which is similar to the simple rectangular pulse as shown in Figure 3.14.

In Figures 3.19 the first null is at $1/B$ (assuming $B\tau >>1$ and an example is shown in Section 3.3.3), which indicates the time delay accuracy is proportional to the sweep bandwidth B instead of the pulse duration τ. This is different from

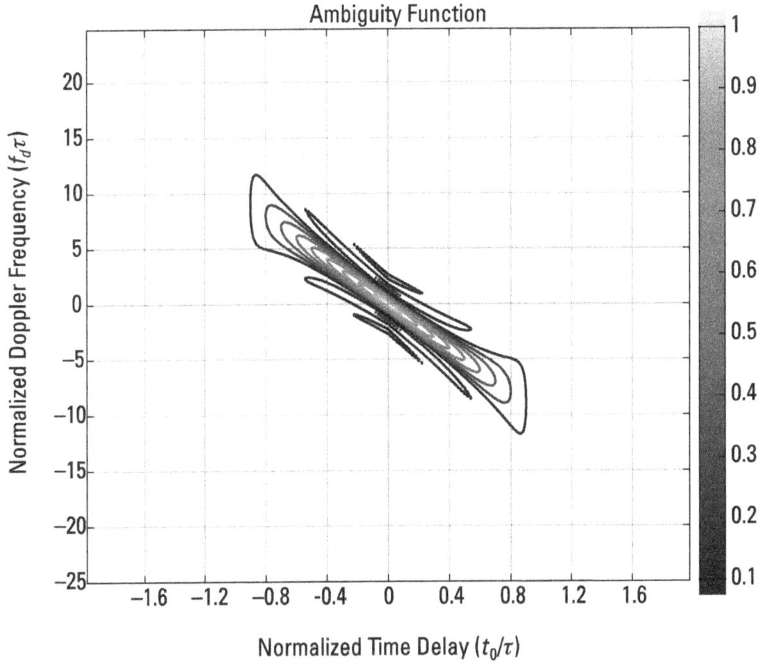

Figure 3.18 Ambiguity function contour of up-chirp LFM.

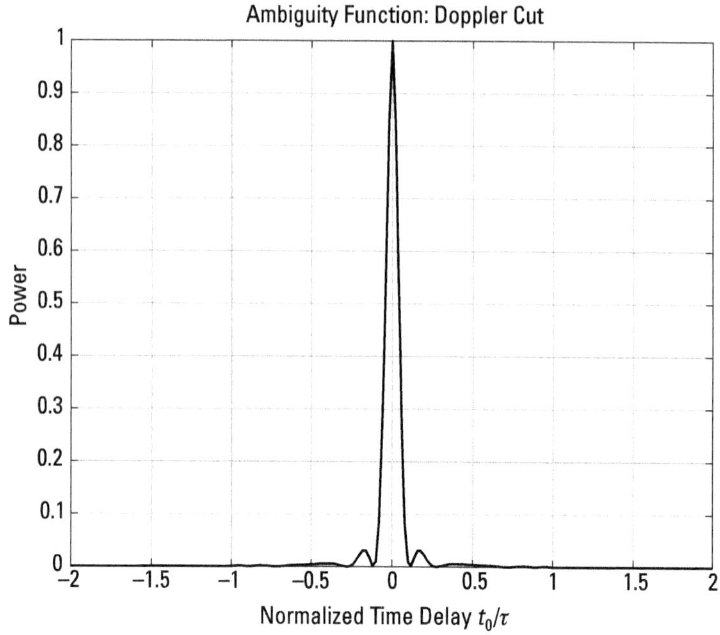

Figure 3.19 Cut along time delay with Doppler frequency $f_d = 0$.

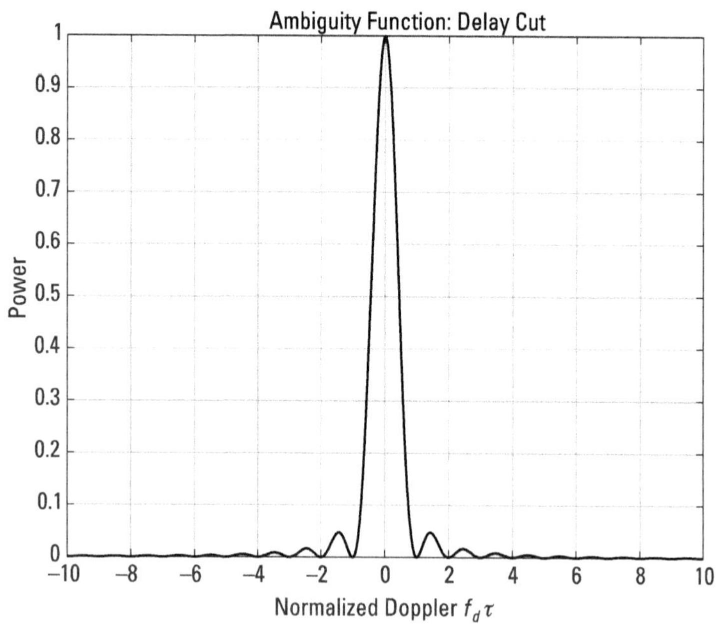

Figure 3.20 Cut along Doppler frequency with time delay $t_0 = 0$.

a rectangular pulse whose time delay accuracy is proportional to τ, as shown in Figure 3.13. The time bandwidth product $B\tau$ can be much greater than 1, and this is an important property that allows radar to utilize longer pulses to achieve better coverage while still being able to obtain good range measurement accuracy and resolution as if a shorter pulse was used. The technique to achieve this is referred to as pulse compression and will be discussed in the following chapter. Unlike the time delay and Doppler frequency accuracies of a simple rectangular pulse, these characteristics of an LFM chirp are independent of each other.

Range Doppler Coupling

The ambiguity function shown in Figure 3.18 is at an angle determined by the sweep slope β. This type of ambiguity function indicates that time delay, Doppler shift, or a combination of both can result in the same output from the matched filter. In other words, a large Doppler shift can result in a range measurement different from the true range. This is known as range Doppler coupling or ambiguity. This can be explained from the ambiguity function of (3.66), where the peak will happen when the following condition is met:

$$\beta t_0 + f_d = 0 \text{ or } t_0 = -f_d / \beta \tag{3.69}$$

When there is no Doppler shift $f_d = 0$, then the peak matched filter output will happen at the expected $t_0 = 0$ moment. However, when there is Doppler shift $f_d \neq 0$, there is Doppler shift resulted range error, which is:

$$\delta R = \frac{-cf_d}{2\beta} \tag{3.70}$$

and c is speed of light. There are both pros and cons associated with range Doppler coupling. It makes LFM chirp more Doppler tolerant compared to a simple unmodulated pulse. Even with input signal having a large Doppler shift, there are still strong matched filter output although at a shifted range from its true location. If we can have signal processing techniques to resolve the range Doppler coupling, there is a single matched filter required that simplifies the system design and cost. The disadvantage is that when there are multiple targets, the signal processing can be tedious and complex.

Range Doppler Decoupling

There are many methods to address the range Doppler coupling to obtain the correct range and Doppler measurement. One method is to send a continuous CW tone after the LFM chirp. From this continuous tone, the Doppler frequency can be measured and the correct range measurement can be obtained through (3.70). This method, however, will have challenges when there are multiple targets in the field of view in that there are multiple Doppler and multiple range measurements to match.

Another common practice [22, 25] is to send a second LFM chirp with different sweeping slope (β_2) following the first LFM chirp (β_1), and an example is shown in Figure 3.21. The target at range R_0 will have a measured range from the first LFM chirp as:

$$R_1 = R_0 + \frac{-cf_d}{2\beta_1} \Rightarrow R_0 = R_1 + \frac{cf_d}{2\beta_1} \tag{3.71}$$

and from the second LFM chirp as:

$$R_2 = R_0 + \frac{-cf_d}{2\beta_2} \Rightarrow R_0 = R_2 + \frac{cf_d}{2\beta_2} \tag{3.72}$$

Since β_1 and β_2 are known design parameters, the unknown R_0 and f_d can be calculated from (3.71) and (3.72) based on the measured R_1 and R_2 values.

Figure 3.22 shows an example R_0-f_d diagram of two target scenarios measured with the triangular up and down chirps. Each line in the figure corresponds to one of the equations for all possible combinations of target range and doppler. The gradient of the line is inversely proportional to the sweep rate of its corresponding equation. The two dotted intersections are respectively indicating the two targets' true range and Doppler values. However, there are two other intersections labeled by square blocks that are not related to any real targets. These are called ghost targets. The up and down chirps (or any two

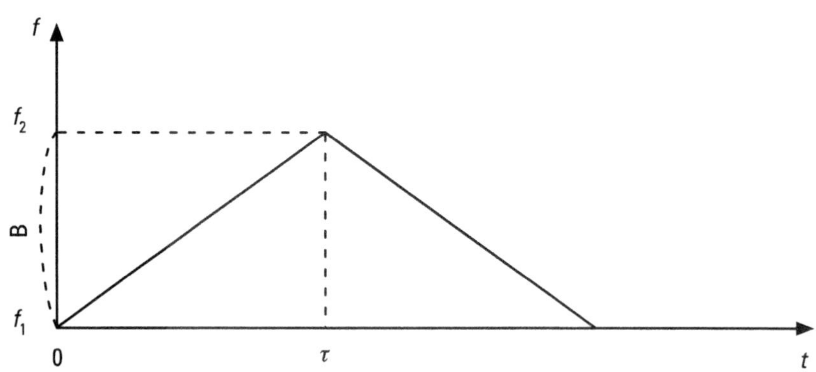

Figure 3.21 Example of FMCW waveforms with triangular modulation. *B* is the sweep bandwidth.

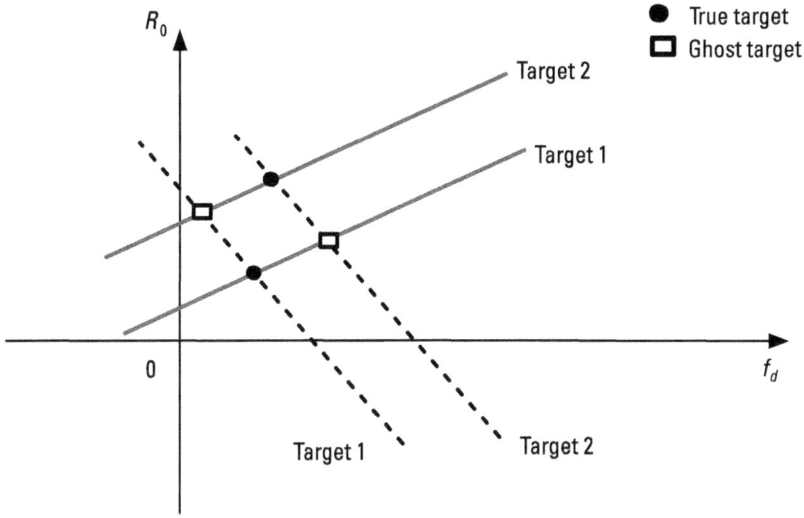

Figure 3.22 Example R_0-f_d diagram of two targets measured with triangular chirp waveform. Solid lines are corresponding to up-chirp and dotted lines are from down-chirp.

chirps with different sweep slopes) work well with single target but will suffer ghost targets when there are multiple targets in the field of view. A common exercise to address ghost targets is to send a third chirp, which has a different sweep slope from the first two chirps. An example is shown in Figure 3.23, and its corresponding R_0-f_d diagram can be viewed in Figure 3.24, where intersections with all three lines are true targets.

The main drawbacks of this type of waveform are that it requires extended measurement time to handle multiple targets, and it is limited by clutters. This is why nowadays more and more radar systems adopt the LFM chirp train waveform as discussed in the next section.

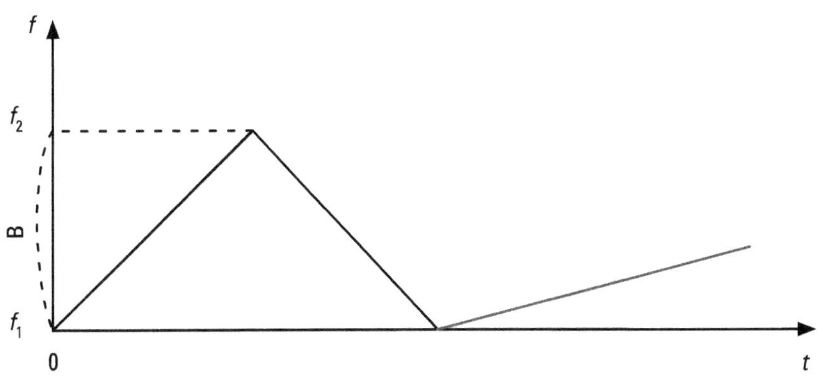

Figure 3.23 Example of FMCW waveforms with triangular modulation.

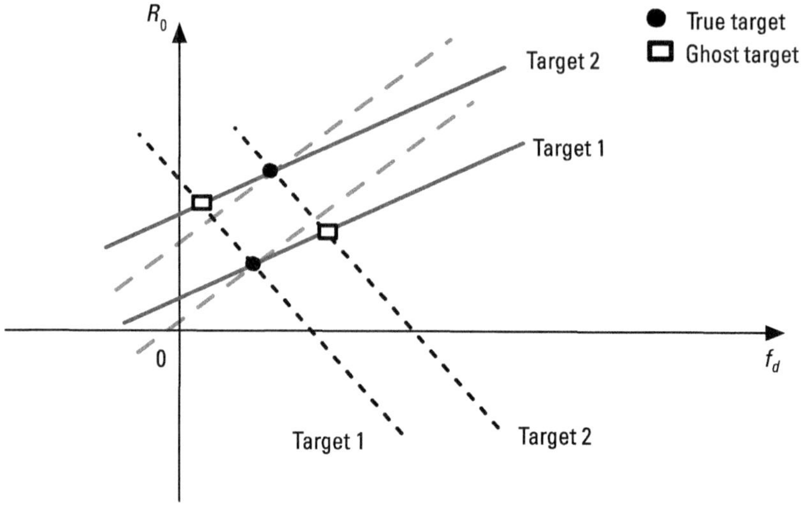

Figure 3.24 Example R_0-f_d diagram of two targets measured with three-chirp waveform. Solid lines are corresponding to up-chirp, dotted lines are from down-chirp, and dashed lines are from the third chirp.

3.3.2 LFM Chirp Train (Fast Chirp)

The LFM chirp train is similar to the traditional LFM chirp except the sweep slope/rate is large. From (3.70) we know that the range measurement error or ambiguity due to Doppler coupling is determined by f_d/β. If we can choose the sweep slope β to be large enough, and for all f_d of interests the range error is much smaller than a range resolution cell (e.g., 10%–50% of the resolution cell), then the coupling effects from Doppler shift can be neglected from consideration during the range processing. The target range can be directly measured from the matched filter output of each individual chirp, and the Doppler frequency can be estimated from a train of chirps. This condition to avoid range Doppler coupling is summarized as the following equation:

$$|\beta| > \frac{\max(f_d) \times c}{2 \times \alpha \times R_{res}} \tag{3.73}$$

where c is speed of light, R_{res} is the range resolution, and α indicates the tolerable range error (e.g., 0.1 means 10% of a range resolution cell).

For automotive long-range radar applications, the maximum relative velocity of interests is less than 100 m/s, which for 77-GHz carrier is corresponding to a Doppler shift:

$$f_d = \frac{2vf_c}{c} = \frac{2\times100\times77\times10^9}{3\times10^8} = 51.33\,kHz \tag{3.74}$$

The typical long-range radar range resolution is around 0.5m. According to (3.73) the sweep rate is required to be $|\beta| > 153.9$ MHz/μs for a conservative $\alpha = 0.1$.

For motion and gesture sensing applications, the maximum relative velocity of interests is less than 10 m/s, which for 60-GHz carrier is corresponding to a Doppler shift:

$$f_d = \frac{2vf}{c} = \frac{2\times10\times60\times10^9}{3\times10^8} = 4\,kHz \tag{3.75}$$

The range resolution is around 0.02m for the 7-GHz available bandwidth in 60-GHz band. According to (3.73) the sweep rate is required to be $|\beta| >$ 300.9 MHz/μs for $\alpha = 0.1$, which is readily achievable nowadays in chip sets as discussed in Chapter 2.

With the fast chirp waveforms, the data processing as discussed in Chapter 4 is straightforward and similar to that of phase coded pulse waveforms. Also, all chirps can be utilized to detect the target of interests, and no chirps are wasted for decoupling, which provides sensitivity advantages over the conventional FMCW waveform. It will be obvious in later chapters that fast chirp modulation can take advantage of the Doppler domain processing to better handle clutters. Due to these benefits and hardware availability, the current and future generation of radars are adopting fast chirp modulations in civil applications.

3.3.3 Stretch Processing

So far, the demodulation of linear FMCW waveforms is discussed with the help of the matched filter. For applications requiring high range resolution, the bandwidth of the waveforms can be a few GHz up to 7 GHz. According to Nyquist sampling theory [26], the ADC needs to have a sampling rate more than twice of the signal bandwidth. This results in high hardware cost or ADCs with limited number of bits and dynamic range. It is worth noting that it is a common practice to add a margin of 20%–50% to the sampling frequency to account for filter transition band and waveform finite bandwidth. For a signal with 7-GHz bandwidth, the minimum sampling frequency is 14 GHz, and in practice it is somewhere between 16.8 GHz and 21 GHz.

If the range of interests with high resolution is limited, there is an alternative to matched filter to demodulate linear FMCW waveforms, which is known as stretch processing [27, 28]. It processes the actual wide bandwidth received signals with relatively much narrower bandwidth receiver circuits. The key idea is to trade time delay for frequency offset. The received signals are mixed with a local oscillator (LO) that ramps at precisely the same rate as the transmitted signal. Very often this LO reference signal is just a replica of the transmitted signal. The output of the mixer will be signals with constant frequencies (beat frequency). The beat frequency will vary based on the relative delay between the received signal and the LO signal. In other words, the return signals from different ranges will have different beat frequencies at the output of the mixer. When the range coverage is reduced (the delay between return signals and the LO is small), the bandwidth of the output signal can be kept within bandwidth of the receiver while the overall range resolution is achieved with the original sweep bandwidth of the transmitted waveform.

For an up-chirp example as shown in Figure 3.25, when the return signal is received, its instantaneous frequency is ramping in parallel to the LO signal and the delta is a constant beat frequency for the overlap of these two signals. The beat frequency is proportional to the range of the return signals as shown in Figure 3.26, and it is obvious that the range is converted to frequency.

The model of stretch processing in complex domain is shown in Figure 3.27, where the LO reference input is a scaled version of the transmitted signal and has the following form:

$$s_c(t) = e^{j\pi\beta t^2} e^{j(2\pi f_c t + \theta_0)} \tag{3.76}$$

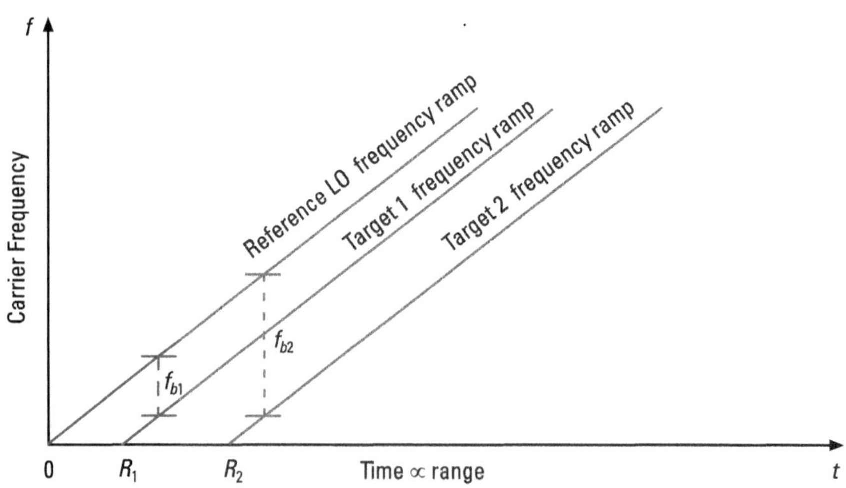

Figure 3.25 Instantaneous frequency ramp of reference LO signal, return signal from target 1, and return signal from target 2.

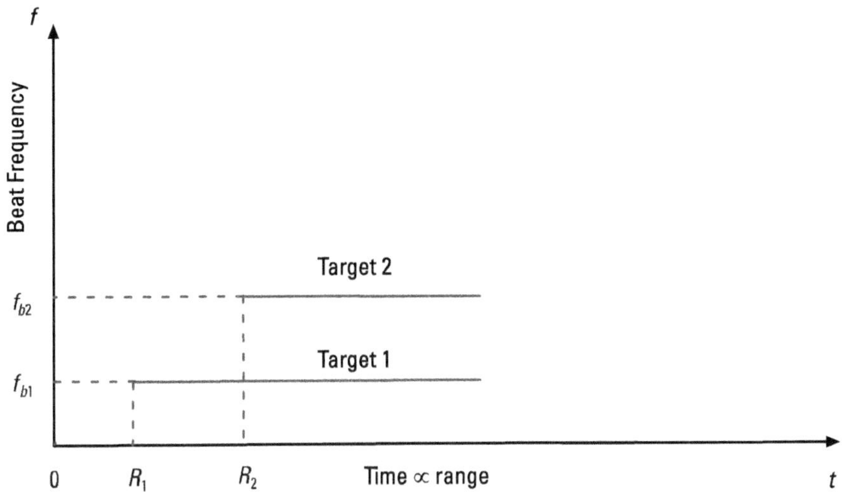

Figure 3.26 The instantaneous frequency of the output signals from the mixer.

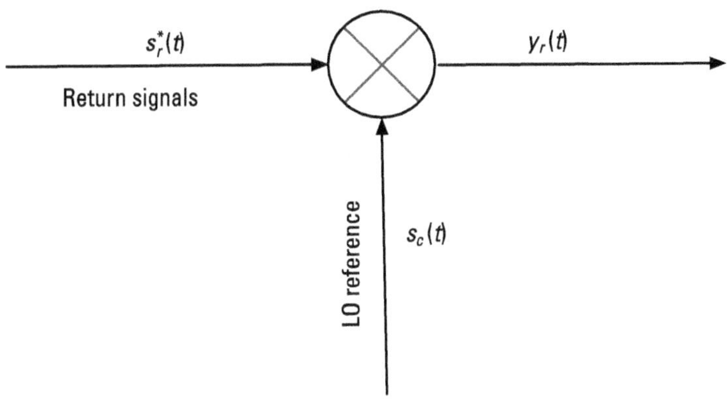

Figure 3.27 Stretch processing complex model.

Let us consider a point target scenario where the return signal has time delay t_0 and Doppler shift f_d, and according to (3.39) and (3.65):

$$s_r(t) = e^{j\pi\beta(t-t_0)^2} e^{j(2\pi(f_c+f_d)(t-t_0)+\theta_0+\phi_0)} \tag{3.77}$$

where the amplitude term is ignored for simplicity of analysis.

The mixer output is:

$$y_r(t) = s_r^*(t)s_c(t) = e^{-j\phi_0} e^{(-j\pi\beta t_0^2 + j2\pi(f_c+f_d)t_0)} e^{j2\pi(\beta t_0 - f_d)t} \tag{3.78}$$

It is obvious that the output signal is a complex sinusoidal with a beat frequency of:

$$f_b = \beta t_0 - f_d \tag{3.79}$$

When f_d is zero, the time delay can be obtained from the measured beat frequency through f_b/β. It is obvious that when $f_d \neq 0$ the range measurement ambiguity is:

$$\delta R = \frac{-cf_d}{2\beta} \tag{3.80}$$

which is consistent with (3.70) and also indicates that the condition (3.73) holds for stretch processing.

The receiver bandwidth is determined by the maximum range R_{max} of interest and the sweep slope:

$$B_r = \beta t_{max} = \frac{Bt_{max}}{T} \tag{3.81}$$

where B is the sweep bandwidth of the transmitted signal, T is the chirp duration, and $t_{max} = 2R_{max}/c$. In civil radar applications where the maximum range of interests is less than 1,000m, the receiver bandwidth is much smaller than the signal's original bandwidth. For example, in motion and gesture sensing applications, the R_{max} is less than 10m. For a 64-μs chirp with 7-GHz sweep bandwidth, the receiver bandwidth is less than 7.3 MHz according to (3.81). The requirement for ADC sampling rate is reduced from 14 GSPS to 14.6 MSPS by about 1,000 times. The time bandwidth product $B\tau$ equals to 448,000 >> 1.

For military radar applications, stretch processing is less favorable in that R_{max} can be large. For the same example if R_{max} is 9 km instead of 10m, then the required receiver bandwidth is about 6.6 GHz, similar to its sweep bandwidth.

3.4 Phase Coded Waveforms

The phase coding is another widely adopted waveform and is finding applications in civil radars such as automotive and gesture sensing. The phase coding can be applied at each pulse but more often at each subpulse of a long waveform as shown in Figure 3.5. The subpulses are called chips or bits, and generally they are of equal length. The long waveform has a constant carrier frequency, and each chip has its own phase modulated according to the codes. The phase coded waveform of N bits can be expressed as:

$$s_c(t) = \sum_{n=0}^{N-1} x_n(t - n\tau_c) e^{j(2\pi f_c t + \theta_0)} = g(t) e^{j(2\pi f_c t + \theta_0)} \tag{3.82}$$

where

$$x_n(t) = e^{j\phi_n} rect\left(\frac{t}{\tau_c}\right) \tag{3.83}$$

f_c is the carrier frequency, θ_0 is an arbitrary initial phase of the carrier, τ_c is the duration of each chip, ϕ_n represents the phase coding, and *rect* is the rectangular function. $g(t)$ is the complex envelope.

The phase changes ϕ_n can be binary (0 or τ radians), and a well-known example is Barker code [29]. Its matched filter output has equal sidelobes (known as time sidelobes) besides the expected main lobe. The Barker code of length N = 13 is shown in Figure 3.28. The matched filter output (autocorrelation function) is shown in Figure 3.29, where there are six equal time sidelobes on either side of the main peak. The Barker codes are the optimal codes in the sense of peak to sidelobe ratio; however, the longest Barker code available is only of length 13.

The selection of the 0 or π phases can be random, which will result in thumbtack-like ambiguity function as shown in Figure 3.9. In practice the random sequence is generated through a shift register with feedback and modulo

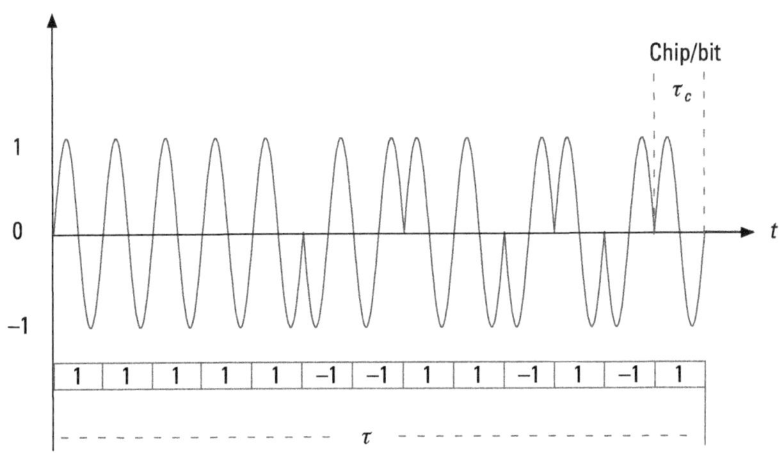

Figure 3.28 Barker coded waveforms when code length = 13.

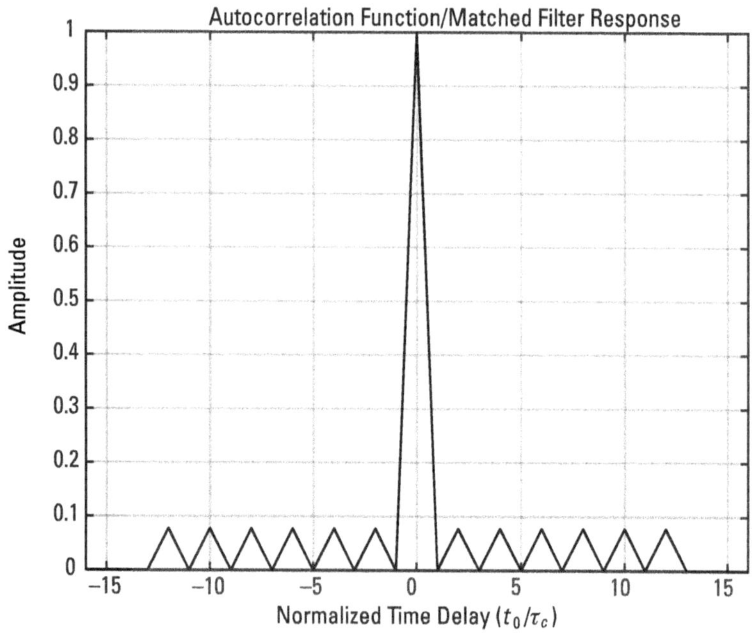

Figure 3.29 Normalized (divided by 13) autocorrelation of 13-bit Barker code.

2 additions [30]. This type of waveform is also known as pseudorandom sequence. A benefit of the pseudorandom sequence is that it can have spread spectrum and provide the capability of rejecting interference as well as allowing multiple simultaneous use of the same frequency band by coding each transmitted signal differently. It should be noted that not all randomly selected codes can provide desired low time sidelobes.

There are two main disadvantages for binary phase coded waveforms. The first one is that they are not Doppler tolerant. When the Doppler frequency of interests is large, it often requires a bank of matched filters with each filter tuned to a different Doppler frequency. The second one is that the time sidelobes can be large with Doppler shift that can result in ghost targets.

The phase changes can be different values other than the binary phases of 0 or π. These are called polyphase codes. Polyphase code can produce lower time sidelobes than the binary phase codes and are better tolerant to Doppler frequency shift. Good examples of polyphase codes are Frank and extended Frank codes [31, 32], which are employed in various real radar systems.

However, nowadays the binary code is attracting more attention in civil radar applications due to its simplicity to implement in hardware. The pseudorandom binary codes are of particular interest in automotive radars due to their capability to accommodate many users in the same frequency band. There are extensive research and planned product releases in automotive radars with binary phase coding, as summarized in [33–35].

3.4.1 Golay Codes

In this section, we discuss a particular type of binary phase code: Golay codes [36], which is utilized as preamble in IEEE802.11 AD/AY protocols [37, 38] and being proposed to motion and gesture sensing radar applications [39–41]. The working principle and demodulations can be applied to other types of phase codes.

The Golay code has a special complementary property, which means there are two equal length pairs whose time sidelobes of the autocorrelation function are exactly the same but at opposite signs. If the autocorrelation functions are added together, the time sidelobes cancel out with zero sidelobes in the final output, and the mainlobe will be $2N$, where N is the number of bits in each of the two codes. These two code pairs are called complementary codes.

On IEEE802.11-AD standard [37] page 490, 128-bit complementary Golay code pairs are adopted and described. The values of the primary Golay code Ga and its complementary pair Gb are listed in Tables 3.1 and 3.2, respectively. The matched filter outputs of Ga and Gb sequences are respectively shown in Figures 3.30 and 3.31. It is obvious there are time sidelobes in these outputs. The sum of these two complementary pairs matched filter output is plotted in Figure 3.32, and all the time sidelobes are reduced to zero. In practice the time sidelobes are not zero even with the complementary pairs due to Doppler frequency shift in the returned signals, as shown in Figure 3.33, which post serious practical challenges. There have been efforts [42] to modify the Golay sequences to be more Doppler tolerant. However, in motion and gesture

Table 3.1

The Primary 128-Bit Golay Sequence $Ga_{128}(n)$

The Sequence $Ga_{128}(n)$, to be transmitted from left to right, up to down
+1 +1 -1 -1 -1 -1 -1 -1 -1 +1 -1 +1 +1 -1 -1 +1 +1 +1 -1 -1 +1 +1 +1 +1 -1 +1 -1 +1 -1 +1 +1 -1
-1 -1 +1 +1 +1 +1 +1 +1 +1 -1 +1 -1 -1 +1 +1 -1 +1 +1 -1 -1 +1 +1 +1 +1 -1 +1 -1 +1 -1 +1 +1 -1
+1 +1 -1 -1 -1 -1 -1 -1 -1 +1 -1 +1 +1 -1 -1 +1 +1 +1 -1 -1 +1 +1 +1 +1 -1 +1 -1 +1 -1 +1 +1 -1
+1 +1 -1 -1 -1 -1 -1 -1 -1 +1 -1 +1 +1 -1 -1 +1 -1 -1 +1 +1 -1 -1 -1 -1 +1 -1 +1 -1 +1 -1 -1 +1

Table 3.2

The Complementary 128-Bit Golay Sequence $Gb_{128}(n)$

The Sequence $Gb_{128}(n)$, to be transmitted from left to right, up to down
-1 -1 +1 +1 +1 +1 +1 +1 +1 -1 +1 -1 -1 +1 +1 -1 -1 -1 +1 +1 -1 -1 -1 -1 +1 -1 +1 -1 +1 -1 -1 +1
+1 +1 -1 -1 -1 -1 -1 -1 -1 +1 -1 +1 +1 -1 -1 +1 -1 -1 +1 +1 -1 -1 -1 -1 +1 -1 +1 -1 +1 -1 -1 +1
+1 +1 -1 -1 -1 -1 -1 -1 -1 +1 -1 +1 +1 -1 -1 +1 +1 +1 -1 -1 +1 +1 +1 +1 -1 +1 -1 +1 -1 +1 +1 -1
+1 +1 -1 -1 -1 -1 -1 -1 -1 +1 -1 +1 +1 -1 -1 +1 -1 -1 +1 +1 -1 -1 -1 -1 +1 -1 +1 -1 +1 -1 -1 +1

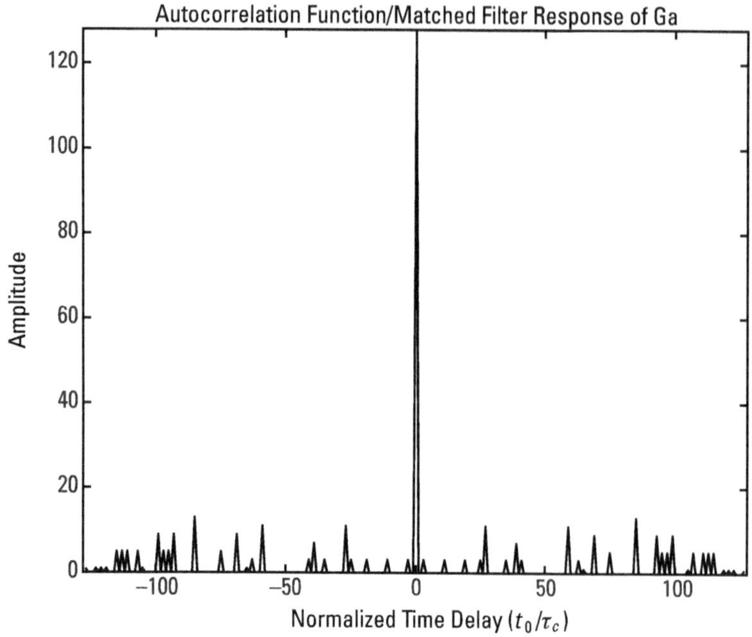

Figure 3.30 The matched filter output of primary Golay 128-bit sequence.

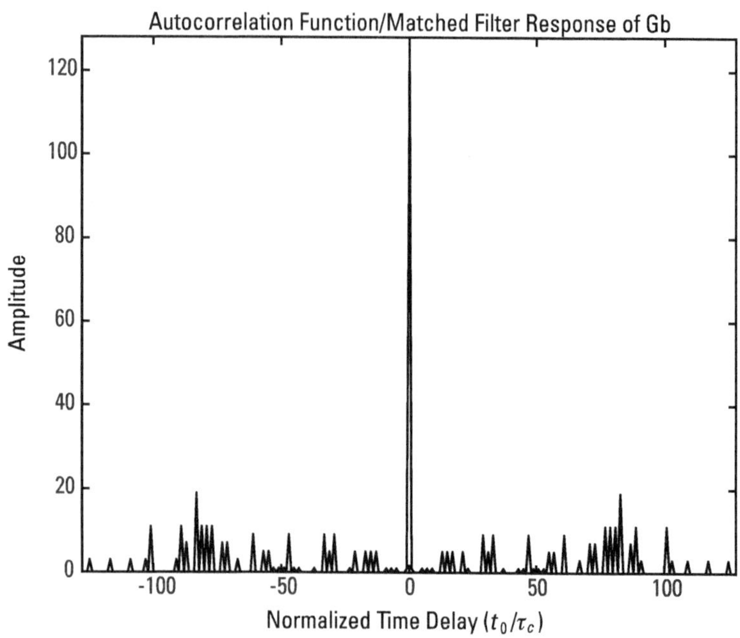

Figure 3.31 The matched filter output of complementary Golay 128-bit sequence.

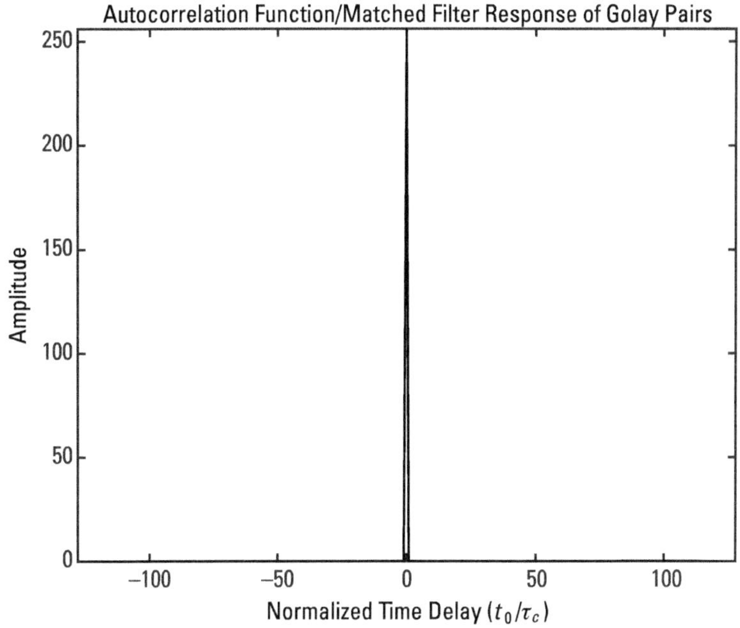

Figure 3.32 The summation of matched filter output of Golay 128-bit pair.

sensing application, the Doppler shift is relatively insignificant and allows the direct application of Golay codes.

The rate for the code to change from one bit to the next is around 2–3 GHz, which requires high-speed ADCs. If each bit lasts 0.46 ns (~2.16-GHz bandwidth), then one Golay sequence is about 58.88 ns. The Golay sequences need to be repeated a few times in order to have enough energy to cover the desired range. The matched filter is implemented right after the ADCs in hardware to reduce the amount of data for post signal processing.

3.5 Summary

The linear FMCW waveform is Doppler tolerate. The stretch processing reduces the receiver bandwidth significantly and results in low system cost. The recent chip development enables fast chirp modulation, which greatly simplifies the signal processing and enhances the radar sensitivity and clutter handling capabilities. Based on these advantages, fast chirp FMCW radars with stretch processing are widely adopted in civil applications such as gesture sensing and automotive radar.

Binary phase coding is in the early stages for automotive and gesture sensing. The main advantage is that it has spread spectrum, which can handle

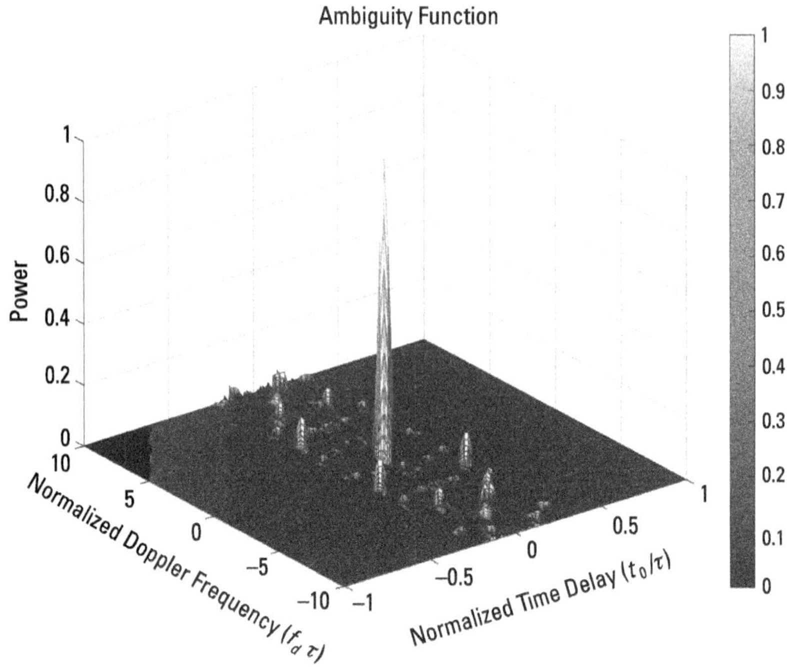

Figure 3.33 A 128-bit complementary Golay pairs ambiguity function diagram. τ is the overall code length.

multiple devices with minimum mutual interference. It also provides the possibility of dual use of the same hardware for communication systems and radar. The challenges of these waveforms are Doppler tolerance and high sampling rate.

References

[1] Peeples, P., *Radar Principles,* Chapter 2, New York: John Wiley & Sons, 2004.

[2] Oswald, J., "The Theory of Analytic Band-Limited Signals Applied to Carrier Systems," *IRE Transactions on Circuit Theory,* Vol. 3, No. 4, 1956, pp. 244–251.

[3] Cook, C. E., and M. Bernfeld, *Radar Signals: An Introduction to Theory and Application*, Norwood, MA: Artech House, 1993.

[4] Richards, M. A., *Fundamentals of Radar Signal Processing,* New York: McGraw-Hill, 2005.

[5] Levanon, N., and E. Mozeson, *Radar Signals,* New Jersey: John Wiley & Sons, 2004.

[6] Rihaczek, A. W., *Principles of High-Resolution Radar,* Norwood, MA: Artech House, 1996.

[7] North, D. O., "An Analysis of the Factors which Determine Signal/Noise Discrimination in Pulsed Carrier Systems," *Proceedings of the IEEE,* Vol. 51, No. 7, 1963, pp. 1016–1027.

[8] Peebles, P. Z., *Probability, Random Variables, and Random Signal Principles*, Fourth Edition, Chapter 8, New York: McGraw-Hill, 2000.

[9] Gradshteyn, I. S., and I. M. Ryzhik, *Tables of Integrals, Series, and Products*, Sixth Edition, San Diego, CA: Academic Press, 2000, p. 1099.

[10] Oppenheim, A. V., and R. W. Schafer, *Discrete-Time Signal Processing*, Second Edition, Upper Saddle River, NJ: Prentice Hall, 1999, p. 60.

[11] Blake, L. V., "A Guide to Basic Pulse-Radar Maximum-Range Calculation, Part 1—Equations, Definitions, and Aids to Calculation," *NRL Report 6930*, December 1969.

[12] Kay, S. M., *Fundamentals of Statistical Signal Processing, Vol. II: Detection Theory*, Upper Saddle River, NJ: Prentice Hall, 1998.

[13] Woodward P. M., *Probability and Information Theory with Applications to Radar*, Oxford: Pergamon Press, 1964.

[14] Temes, C. L., "Relativistic Consideration of Doppler Shift," *IRE Trans. on Aeronautical and Navigational Electronics*, Vol. ANE-6, No. 1, 1959, p. 37.

[15] Morin, D., *Introduction to Classical Mechanics: With Problems and Solutions*, Chapter 11, Cambridge, UK: Cambridge University Press, 2007.

[16] Stimson, G. W., et al., *Introduction to Airborne Radar*, Third Edition, Chapter 18, Edison, NJ: SciTech Publishing, 2014.

[17] Peeples, P., *Radar Principles*, Chapter 1.6, New York: John Wiley & Sons, 2004.

[18] Siebert, W. M., "A Radar Detection Philosophy," *IRE Trans. IT-2*, pp. 204–221.

[19] Arfken, G., *Mathematical Methods for Physicists*, Third Edition, Orlando, FL: Academic Press, 1985.

[20] Skolnik, M. I., *Introduction to Radar Systems*, Third Edition, Chapter 6.4, New York: McGraw-Hill, 2001.

[21] Lutz, S., et al., "On Fast Chirp Modulations and Compressed Sensing for Automotive Radar," *International Radar Symposium*, 2014, pp. 1–6.

[22] Stove, A. G., "Linear FMCW Radar Techniques," *IEE Proceedings F - Radar and Signal Processing*, Vol. 139, No. 5, 1992, pp. 343–350.

[23] Winkler, V., "Range Doppler Detection for Automotive FMCW Radars," *European Radar Conference*, 2007, pp. 1445–1448.

[24] Patole, S. M., et al., "Automotive Radars: A Review of Signal Processing Techniques," *IEEE Signal Processing Magazine*, Vol. 34, No. 2, 2017, pp. 22–35.

[25] Rohling, H., and M. Meinecke, "Waveform Design Principles for Automotive Radar Systems," *CIE International Conference on Radar Proceedings*, Beijing, China, 2001, pp. 1–4.

[26] Oppenheim, et al., "Discrete-Time Signal Processing," Second Edition, Chapter 4, Upper Saddle River, NJ: Prentice-Hall, 1989.

[27] Caputi, W. J., "A Technique for the Time-Transformation of Signals and Its Application to Directional Systems," *Radio and Electronic Engineer*, Vol. 29, No. 3, 1965, pp. 135–141.

[28] Caputi, W. J., "Stretch: A Time-Transformation Technique," *IEEE Transactions on Aerospace and Electronic Systems*, Vol. AES-7, No. 2, 1971, pp. 269–278.

[29] Barker, R. H., *Group Synchronizing of Binary Digital Systems, Communication Theory*, London: Butterworth, 1953, pp. 273–287.

[30] Skolnik, M. I., *Radar Handbook*, Third Edition, Chapter 8.2, New York: McGraw-Hill, 2008.

[31] Frank, R., "Polyphase Codes with Good Nonperiodic Correlation Properties," *IEEE Transactions on Information Theory*, Vol. 9, No. 1, 1963, pp. 43–45.

[32] Lewis, B. L., and F. F. Kretschmer, "A New Class of Polyphase Pulse Compression Codes and Techniques," *IEEE Transactions on Aerospace and Electronic Systems*, Vol. AES-17, 1981, pp. 364–371.

[33] Bourdoux, A., et al., "PMCW Waveform and MIMO Technique for a 79 GHz CMOS Automotive Radar," *IEEE Radar Conference*, Philadelphia, PA, 2016, pp. 1–5.

[34] Hakobyan, G., and B. Yang, "High-Performance Automotive Radar: A Review of Signal Processing Algorithms and Modulation Schemes," *IEEE Signal Processing Magazine*, Vol. 36, No. 5, 2019, pp. 32–44.

[35] Alland, S., et al., "Interference in Automotive Radar Systems: Characteristics, Mitigation Techniques, and Current and Future Research," *IEEE Signal Processing Magazine*, Vol. 36, No. 5, 2019, pp. 45–59.

[36] Golay, M., "Complementary Series," *IRE Trans. Inform. Theory*, Vol. 7, No. 2, 1961 pp. 82–87.

[37] IEEE Standard for Information Technology—Telecommunications and Information Exchange Between Systems—Local and Metropolitan Area Networks—Specific Requirements-Part 11: Wireless LAN Medium Access Control (MAC) and Physical Layer (PHY) Specifications Amendment 3: Enhancements for Very High Throughput in the 60 GHz Band," IEEE Std 802.11ad-2012, 2012.

[38] Ghasempour, Y., et al., "IEEE 802.11ay: Next-Generation 60 GHz Communication for 100 Gb/s Wi-Fi," *IEEE Communications Magazine*, Vol. 55, No. 12, 2017, pp. 186–192.

[39] Grossi, E., et al., "Opportunistic Radar in IEEE 802.11ad Networks," *IEEE Transactions on Signal Processing*, Vol. 66, No. 9, 2018, pp. 2441–2454.

[40] Kumari, P, et al., "IEEE 802.11ad-Based Radar: An Approach to Joint Vehicular Communication-Radar System," *IEEE Transactions on Vehicular Technology*, Vol. 67, No. 4, 2018, pp. 3012–3027.

[41] Hof, E., et al., "Gesture Recognition with 60GHz 802.11 Waveforms," *International Conference on Artificial Intelligence in Information and Communication (ICAIIC)*, 2020, pp. 325–329.

[42] Pezeshki, A., et al., "Doppler Resilient Golay Complementary Waveforms," *IEEE Transactions on Information Theory*, Vol. 54, No. 9, 2008, pp. 4254–4266.

4

Radar Signal Processing

After being received and down-converted in the receiver, the radar return signals are usually digitized and then subject to further signal and data processing to extract the desired information such as the target's range, Doppler, angle, reflectivity, and so on. The signal processing is generally referring to detect desired return signals and reject noise, interference, and clutter (undesired echoes). It may consist of different components for different radar applications. In this chapter, we will focus on the signal processing chain for motion and gesture detection radar; however, these techniques are generally applicable to many other civil applications such as automotive radar. The data processing is traditionally the process of acquiring further information about the target such as location, trajectory, and recognition/classification. For motion and gesture radar, the data processing is about understanding the behavior and motion pattern of the target, which benefits more from modern artificial intelligence (AI) technologies.

A typical signal and data processing flow is shown in Figure 4.1. The received signal after digitization will normally have range processing or range gating to separate targets into different range bins. This processing is often combined with pulse compression (to improve the range resolution) and matched filter (to improve the signal-to-noise ratio). At each range bin and across multiple waveforms, the signal is subject to Doppler processing to separate the targets into different Doppler bins and away from clutter. There is often a need to have clutter cancellation or DC leakage suppression before the Doppler processing as part of the signal conditioning. This is particularly important for continuous wave operation in that the receiver is not gated off during transmission and there are many strong return signals from surrounding infrastructures and internal leakage. The results of range and Doppler processed data form 2-D range Doppler spectrum. These data are sometimes processed with specially

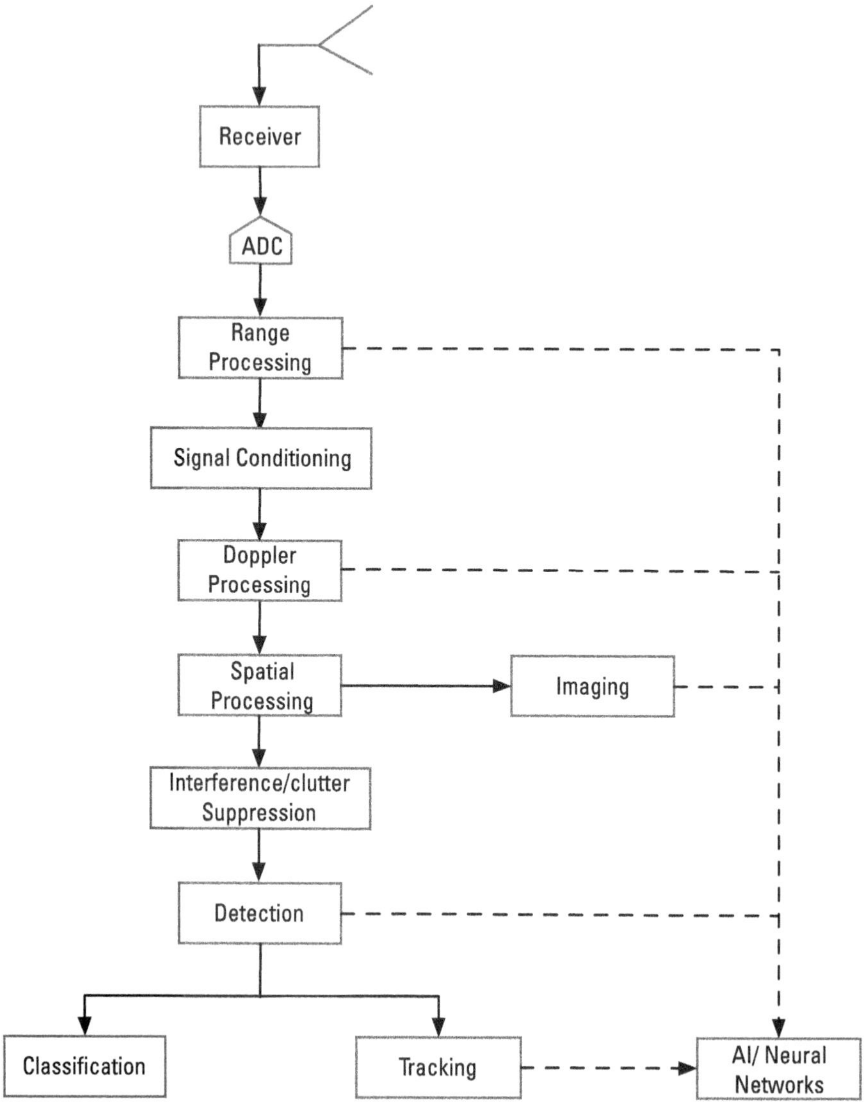

Figure 4.1 A typical radar signal data processing flow.

designed filters to suppress clutter or interference prior to further beamforming and angle estimation. The order of these signal processing blocks is not necessarily fixed due to the linear system nature. The range and Doppler processing can be considered as integrations to improve the signal-to-noise ratio prior to detection. After detection, there can be high-level tracking and classifications. For motion and gesture sensing, the effective classification methods are of significant importance. There are conventional methods such as nearest neighbor or random forest; however, the state of the art is based on deep learning neural networks, which demonstrate better recall rate and precision.

In this chapter we will focus on range processing and Doppler processing. Spatial processing is discussed in Chapter 5. Detection and clutter/interference filtering will be addressed in Chapter 6. High-level classification and recognition are explained in Chapter 7.

4.1 Range Processing (Fast Time Processing)

The range processing is usually applied to data within a range of hundreds of nanoseconds to hundreds of microseconds, and it is often referred to as fast time processing. For a simple pulsed waveform, each pulse lasts only a few hundred nanoseconds to achieve the desired range resolution for gesture detection. For FMCW waveforms, each chirp is about a few dozen microseconds with a large sweep frequency bandwidth. The range processing theories of these waveforms are discussed in Sections 3.3.3 and 3.4.1, respectively, for chirp and pulsed waveforms.

In practice, stretch processing is generally utilized for FMCW waveform due to its simplicity. The fast Fourier transform (FFT) is applied to the digitized signal at the output of ADC, and targets at different ranges will show up at different (beat) frequency bins/range gates. The size of each range gate is determined by:

$$\Delta R = \frac{D_{size} c}{2\beta} \tag{4.1}$$

where D_{size} is the size of each Doppler bin in Hz, c is speed of light, and β is sweep slope in Hz/s. Note range gate size simply indicates what each range gate represents in physical meters, which is not necessarily the same as range resolution, as discussed in the following section. One can oversample the data to have very fine range gate size, which, however, will not change the range resolution.

For phase coded waveforms as discussed in Section 3.4, the matched filter or its correlator form [1] is often implemented in hardware to address the high data rate. For gesture detection application, the bandwidth of the signal is a few gigahertz, which means the ADC has to run at higher rate to satisfy Nyquist theorem (> twice the bandwidth). The signal output of the correlator can be reduced by keeping the relevant range cells only, which can lower the data rate by a few orders of magnitude. The final sampling time Δt determines the range gate's size:

$$\Delta R = \frac{\Delta tc}{2} \tag{4.2}$$

4.1.1 Minimum Range and Maximum Unambiguous Range

The radar systems discussed in this book are capable of keeping the receiver open while transmitting. This enables the detection of these targets that are very close to radar. Technically, the minimum range is zero meters; however, in practice the first range gate is usually heavily contaminated with DC leakage, and other nearfield clutter/interference and will be discarded.

The maximum unambiguous range is determined by the pulse repetition interval (PRI) for pulse waveform:

$$R_{max} = \frac{PRI \times c}{2} \tag{4.3}$$

For LFM system with stretch processing, the maximum range is often constrained by the maximum tolerable signal loss. As shown in Figure 3.26, the targets from further range will have shorter duration after stretch processing in that their starting time is proportional to the round trip delay. In practice, the rule of thumb is 10%–20% of the chirp length τ_c. For a 32 μs chirp, if 10% of signal loss is allowed, the maximum range is:

$$R_{max} = \frac{0.1\tau_c \times c}{2} = 480m \tag{4.4}$$

4.1.2 Pulse Compression

In most radar systems, both range resolution and detection range are important figures of merit. High range resolution may be obtained by a simple (non-modulated) pulse with short duration. Since the sensitivity of a radar relies on the energy contained in the transmitted pulse as discussed in Chapter 2, long detection range requires a long duration pulse and/or high peak power. Simply increasing the radar pulse length has the side effect of degrading the range resolution. Alternatively, short pulses require higher transmitter power, which translates to higher cost and lower reliability. Pulse compression techniques [1–3] allow reduced transmitter power while still maintaining long detection range and high resolution. The reason is that the range resolution does not necessarily depend on the duration of the waveform but on the bandwidth of the waveform. The transmitted long pulses can be modulated to increase their bandwidth, and on receive these pulses are compressed in the time domain, resulting in a range resolution higher than that associated with a nonmodulated pulse.

The operation of pulse compression can be described with a simple short pulse and a long-modulated pulse. Let us assume the simple short pulse has a width of τ and then its associated spectral bandwidth $B \approx 1/\tau$. When the long pulse of width $T >> \tau$ is used, it can be modulated to have a spectral bandwidth much bigger than $1/T$. In fact, it can be modulated to have the same bandwidth B as that of the short pulse, and the resulting time bandwidth product (BT) is much bigger than 1. The modulation is generally in frequency or phase. Amplitude modulation is rarely used due to its low transmit efficiency. On receive, the modulated long pulse return signal is processed in the matched filter to obtain a compressed pulse of width τ at the output. Its range resolution can be found from its ambiguity diagram. For linear FM chirp, its ambiguity diagram as shown in Figure 3.19 indicates the compressed pulse/chirp width equal to the $1/B$, where B is the chirp span bandwidth. For the phase coded pulse, the compressed pulse width equals to its subpulse (bit) width as shown in Figures 3.29 and 3.32. The phase coded long pulse has similar spectral bandwidth to that of its subpulse.

The pulse compression ratio is defined as T/τ the ratio of the long pulse width to the compressed pulse width, which can also be expressed as time frequency bandwidth BT. The pulse compression ratio in practical radar system can be as low as 13 or greater than 100,000. For motion and gesture application, the ratio is about a few thousand for phase coding waveform and around 10^5 for LFM.

Pulse compression is an important concept that in practice is naturally realized through matched filter operation without extra processing. Another way to understand this concept is that the long pulse is marked with different frequency (such as LFM) or phase changes (such as phased coded waveforms) along the various portion of the waveform. On receive, the uniquely marked sections are separated, realigned, and added up by the matched filter. This effectively compresses the received long pulse signal. This can be readily visualized in the example of Figure 4.2, where the binary phase coded waveform is compressed by the delay line filter.

4.1.3 Range Resolution

Resolution is the capability to revolve close spaced point targets or scatterers. Range resolution is the minimum distance at which two equal RCS scatterers can be separated and discerned as individual scatterers. Strictly speaking, range resolution is also associated with a likelihood to indicate how likely the two scatterers can be resolved due to the stochastic nature of radar signals. These

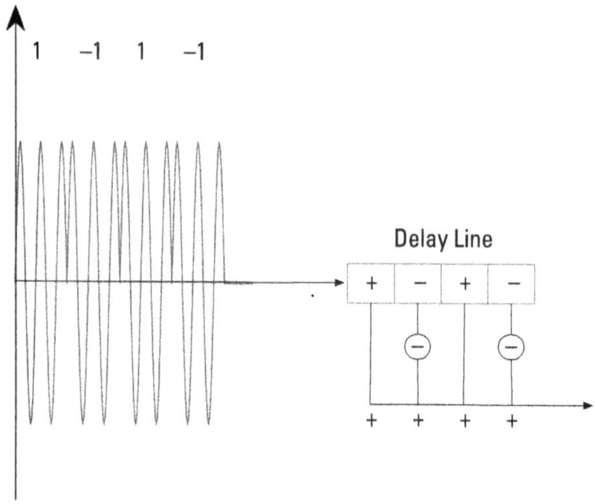

Figure 4.2 Received phase coded signal is compressed by a tapped delay line filter. Plus sign means 0° phase, negative sign means 180° phase, and circle with negative sign in the middle represents 180° phase shift.

scatterers might be individual targets such as two people, or might be body parts of a single person such as fingers, hand, arm, or head. For example, if a radar system has a range resolution of 0.5m with 0.8 probability, it can resolve two equal scatterers at 0.5m apart with 80% of the time. The probability of range resolution is defined as the probability with which the radar reports distinct detection for each of the closely spaced targets and the corresponding range estimation errors are less than a predefined error bound. In practice, it is evaluated through averaging the detection results over a large number of trials, and in each trial the relative phase of the closely spaced targets is randomly varied. It should be pointed out that when the two targets have different RCS, the range resolution is generally worse. For extreme cases (one is extremely strong and one is extremely weak), the radar may never resolve these scatterers regardless of their separation distance. When discussing resolutions, we will assume that the scatterers have the same RCS with equal amplitude return signals, and the return signals are strong enough to be detected.

The overall range resolution is affected by many factors, including *hardware related*, such as waveform bandwidth and sampling, and *signal processing related*, such as matched filtering, integration, detection, and post processing. A wide range of signal processing techniques can be applied, and their resulting resolution varies dramatically; however, the matched filter is fixed and determined by the transmitted waveform. This is the reason that range resolution is generally defined at the output of matched filter with certain criterion to avoid ambiguity and confusion. There are different criteria in the literature or in practice, and the most often referred ones are pulse width (at half amplitude), 0.8 pulse width (rule of thumb according to [4]), and Rayleigh criterion [5], where the pulse width is from peak to the first zero. Note the pulse width is that of a compressed pulse since it is defined at the output of matched filter.

These criteria can be studied with a simple rectangle pulse of length τ (Figure 4.3(a)) at transmit to demonstrate their relationship. The individual matched filter outputs of the received echoes from two equal RCS scatterers are shown in Figure 4.3(b). Let us assume the first echo is from distance $R_0 = ct_0/2$ and the second echo is from distance $R_1 = c(t_0 + \tau)/2$ with c being the speed of light. Thereafter, the receive time difference of the two echoes is τ, which corresponds to distance separation of $c\tau/2$. The matched filter outputs of these echoes have equal amplitude; however, their phases are not necessarily the same. Instead these outputs can have any relative phase difference in that the phases are related to distance (half wavelength = 360° for two-way range) and phase modulation during reflection. The combined matched filter outputs of these echoes are shown in Figure 4.4. If the matched filter outputs of these two echoes are in phase, the overall outputs are shown in Figure 4.4(a), where the two outputs add up coherently to form a trapezoid. If these outputs are out of phase (180° phase delta), the overall matched filter outputs are shown in Figure 4.4(b), where the two outputs cancel each other to have a deep null at the middle. Let us assume the criterion to resolve two scatterers is to have any valley between the peaks of the two targets, and a more rigorous mathematical treatment based on generalized likelihood-ratio test can be found in [6]. Then the two scatterers can be resolved as long as there is a nonzero phase delta at the filtered outputs. If the distance is more than $c\tau/2$, the two scatterers will be guaranteed to be resolved. For this example, Rayleigh resolution and pulse width resolution are both equal to $c\tau/2$, which is the same as that of a simple pulse without matched filter. This indicates that the matched filter has no negative impact on range resolution. The associated probability of resolution is almost 100% with the only exception that the two echoes are exactly in phase. The rule of thumb resolution is $0.8c\tau/2$ with a probability of ~0.9 [6].

For most radar applications the product $B\tau$ is approximately 1 between the 3 dB power (half power) bandwidth B at transmission and 3 dB amplitude (half amplitude) width τ of compressed pulse width. Thereafter in radar literatures and product specifications the range resolution is generally referred to be:

$$R_{res} = c/2B \tag{4.5}$$

instead of $c\tau/2$. This expression avoids the ambiguity on the compressed or original pulse width. It also applies to LFM waveform as discussed in Section 3.3 with the bandwidth B being sweep bandwidth (3.59). The reason is that after stretch processing the range resolution is equivalent to beat frequency reso-

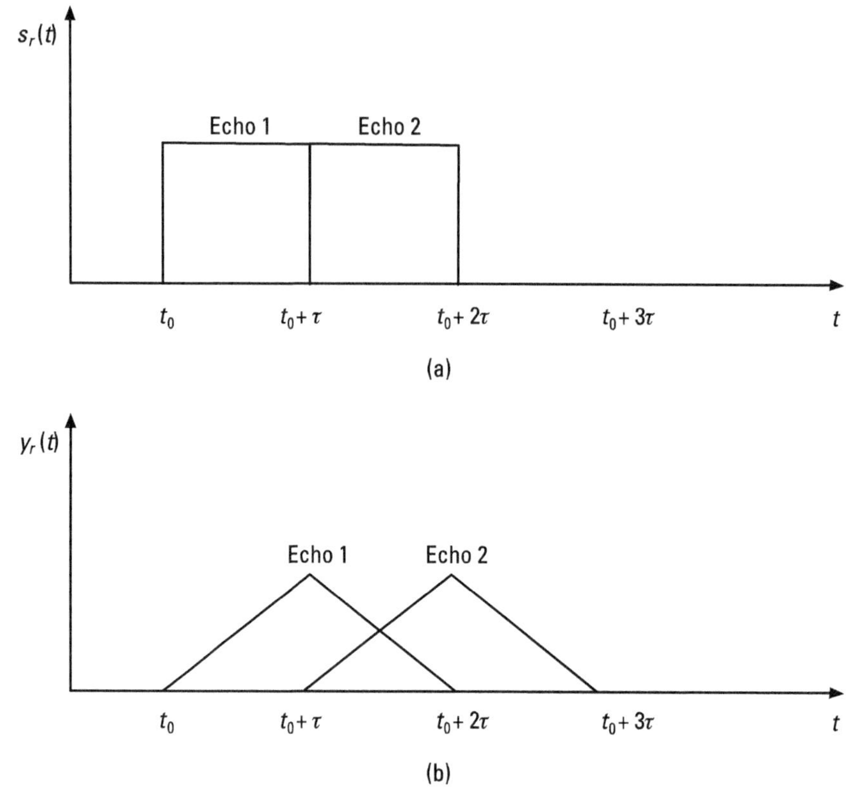

Figure 4.3 (a) Envelope of the two received echo signals at the input of matched filter. (b) Envelope of individual matched filter outputs of received echo signals.

lution. The beat frequency resolution is the inverse of the overall chirp length τ and the frequency sweep rate is B/τ, which mean the two way ranging time of a frequency resolution cell is:

$$t = \frac{1/\tau}{B/\tau} = \frac{1}{B} \tag{4.6}$$

which corresponds to a distance of $c/2B$.

Note the overall system level resolution is also subject to the high level signal processing techniques and may be different from what we have discussed and defined so far. For example, detectors could degrade range resolution considerably [7]. However the metric (4.5) provides a common ground to evaluate the basic hardware capability of different radar systems.

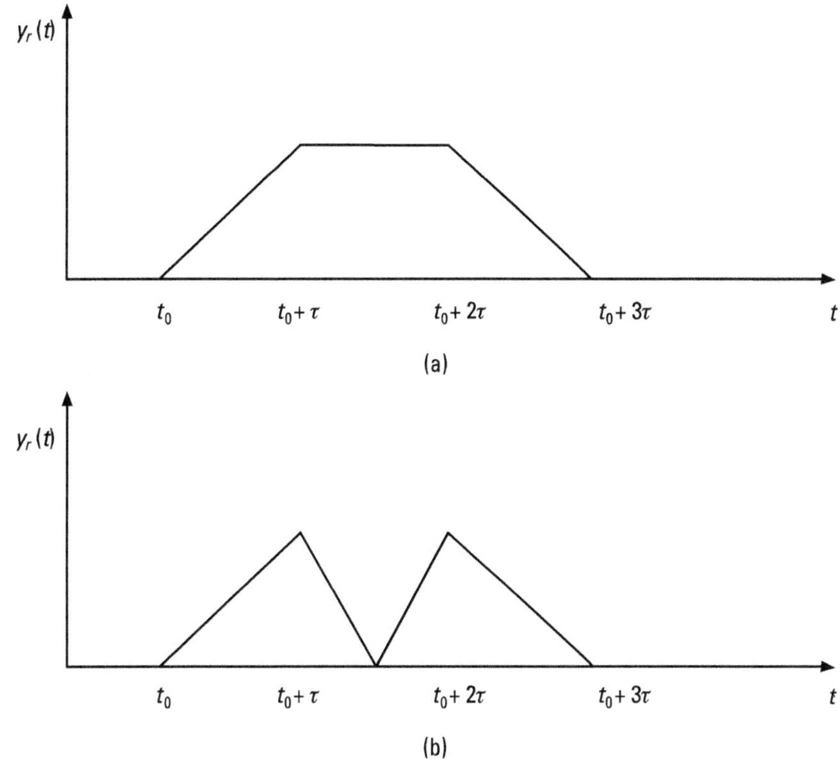

Figure 4.4 (a) Combined matched filter outputs for echoes with the same phase. (b) Combined matched filter outputs for echoes with phase difference of 180°.

4.1.4 Range Accuracy

Accurate extraction of information from radar signal is fundamentally limited by noise. The metrics to evaluate how good this measurement information is accuracy (measurement error), which is defined as the root mean square (RMS) of the difference between the measured or estimated value and the true value. The accuracy indicates how close the measured value is to the true value at a predefined probability. When the accuracy is discussed, it is generally assumed that the signal-to-noise ratio is high (detection is guaranteed) and noise is the only contributor; other errors such as bias are not included. The bias error associated with carefully designed calibration and measurement process is generally small relative to measurement error due to random noise [8]. These assumptions apply to all accuracy discussions in this book: range, Doppler, and angle.

The range accuracy δR is tied to the time delay accuracy δT in that the measurement of range is through the round-trip travel time from radar to target and back:

$$\delta R = \frac{\delta T \times c}{2} \tag{4.7}$$

Various statistical analyses [9, 10] all lead to the following results for range accuracy:

$$\delta T = \frac{1}{B_{eff}\left(2E / N_0\right)^{1/2}} \tag{4.8}$$

where E is the signal energy, N_0 is the noise power per unit bandwidth, and B_{eff} is called the effective bandwidth and defined as [11]:

$$B_{eff}^2 = \frac{1}{E}\int_{-\infty}^{\infty} (2\pi f)^2 \left|S(f)\right|^2 df \tag{4.9}$$

where $S(f)$ is the signal envelope spectrum, and B_{eff}^2 is the normalized second moment of the spectrum about the mean at $f = 0$. It is obvious that the range accuracy is inversely proportional to the effective bandwidth and square root of signal-to-noise ratio. Another interesting observation is that the more spectral energy concentrates at the two ends of the band, the larger the effective bandwidth with better accuracy. This is in contrast to range resolution, which is only affected by the half-power bandwidth and not the shape of the signal spectrum.

The value of B_{eff} is determined by the shape of the signal envelope spectrum and the finite rise and fall time of the pulse. Various examples are shown in [11], and for a simple pulse with half-amplitude width of τ the range accuracy is:

$$\delta R = \frac{\tau \times c}{2 \times 2.1 \times \left(2E / N_0\right)^{1/2}} \tag{4.10}$$

Consider that $B\tau \approx 1$ and (4.5), the range accuracy can be expressed as:

$$\delta R \approx \frac{c}{2 \times 2.1 \times B\left(2E / N_0\right)^{1/2}} \approx \frac{R_{res}}{2.1 \times \left(2E / N_0\right)^{1/2}} \tag{4.11}$$

It is clear that the range accuracy depends on range resolution and the signal-to-noise ratio. As a rule of thumb, range accuracy can be assumed to equal to 10% of range resolution, which is a 10-dB signal-to-noise ratio according to (4.11).

4.1.5 Time Sidelobes Control

When pulse compression is utilized, the matched filter output has sidelobes besides the main target peak. Although this is optimal for a single target scenario, there might be issues when multiple targets and clutter exist. These sidelobes are time/range offset from the main target peak, and if their level is high, it can result in false detections or mask the weak targets at different ranges. These sidelobes are referred to as range sidelobes or time sidelobes. The root cause of these sidelobes are due to discontinuity in the waveforms. This manifests in the finite length for LFM waveform, and in the phase jumps among subcodes in phase coded waveforms. Various techniques have been proposed to minimize the sidelobes, and in this chapter we will only focus on the waveforms suitable for civil applications.

FM Waveforms

It is well known that the basic LFM waveform has a range sidelobe of −13.2 dB with respect to the main lobe after matched filter. This can be readily explained from the stretch processing output as shown in (3.73), where the mixer output is a complex sinusoidal with finite length and its spectrum is sinc function with sidelobes of −13.2 dB [12].

There are two types of methods to reduce the sidelobes of LFM waveforms. The most common one is to apply weighting/windowing to the amplitude of the received signal [13] prior to Fourier transform, which fundamentally forms a combined mismatch filter such that the output signal spectrum has a taper that reduces the sidelobes. In fact, this problem is analogous to spectral leakage in Fourier analysis of a function with finite length. The windows such as Hamming, Blackman, Dolph-Chebyshev, Taylor, and many others [14] used for spectrum leakage control are directly applicable here. This method is straightforward to apply and becomes a common practice in radar applications with LFM waveforms including automotive and gesture control radar. An embodiment is shown in Figure 4.5 as an example. The main tradeoffs are broader mainlobe and signal loss in the processing gain, since the optimal matched filter has been replaced by a mismatched filter. The signal loss will affect the final signal-to-noise ratio and detection performance, and the mainlobe increase means degraded range resolution and accuracy. When window function is applied, the range resolution formula (4.5) changes to $R_{res} = \beta_{win} c/2B$ with $\beta_{win} = 1$ for no window or rectangular window and $\beta_{win} > 1$ for other window functions. When deciding on the window function of choice, the following steps should be considered:

- It is important to understand sidelobe requirements in detail, such as equal sidelobe peaks, slow roll off, or fast roll off of sidelobe peaks. It is

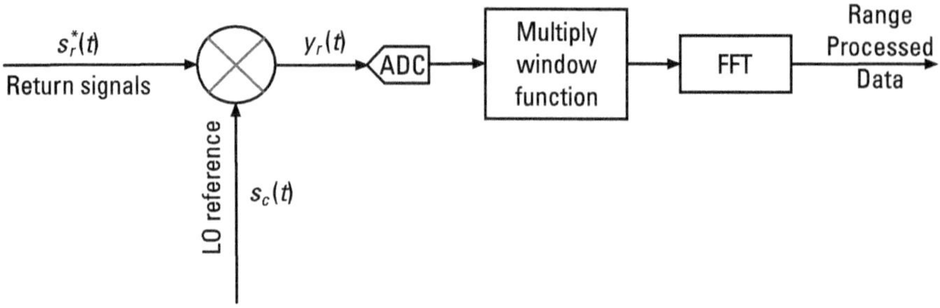

Figure 4.5 Range processing flow for LFM waveforms with windowing to control sidelobes.

not necessarily good to use window function with sidelobe level much more aggressively than required. The achievable sidelobes level can be limited by practical constraints such as hardware imperfections. Overaggressive window function may result in unnecessary signal loss and mainlobe widening. In civil applications, a well-designed radar should target 30–40 dB sidelobes.

- Among the window functions that meet sidelobe requirements, the choice is made based on the priority of either signal loss or the mainlobe width. Taylor window is a popular candidate in that it has the minimum mainlobe width widening [15] and relatively small signal loss for a specified sidelobe level.
- Sometimes the computation considerations determine the window selection. For example, the Hanning window can be implemented in frequency domain with addition and shift only operations, which is ideal for ASIC or FPGA implementation [16].

Example sidelobe levels are shown in Figure 4.6 for no window, Hamming window, and Taylor window of –43 dB sidelobe level. The sidelobes are greatly reduced for Hamming and Taylor windows at the price of wider mainlobe width and lower mainlobe peak. It should be noted that the signal loss is about 5 dB; however, the overall SNR loss is half of that (about 2.5 dB) in that the noise power is also reduced by the window function. It can also be observed that the mainlobe of Taylor window is slightly narrower than that of Hamming window.

The range sidelobes can also be reduced by a nonlinear-frequency-modulation (NLFM) law. For example, the sweep slope of the waveform can vary inversely with the square root of the Taylor window function (higher rate at both ends and lower rate near the center of the waveform), and the matched filter output has similar sidelobe level to that of Taylor window weighted LFM [15, 17]. This type of waveform is widely adopted by air traffic control radars

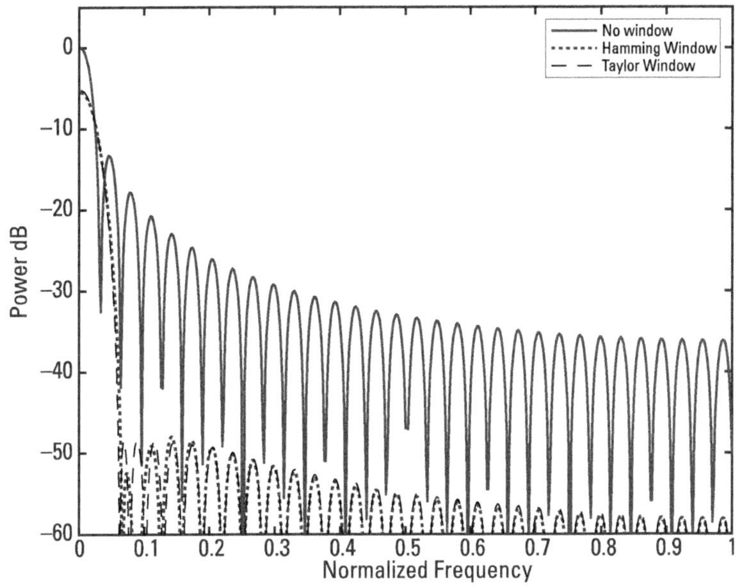

Figure 4.6 No window (rectangular window): solid trace with first sidelobe at −13.2 dB; Hamming window: dotted trace with mainlobe at −5.4 dB and sidelobes below −48 dB (−43 dB sidelobe level relative to mainlobe); Taylor window: dashed trace with mainlobe at −5.2 dB and sidelobes below −48 dB (−43 dB sidelobe level relative to mainlobe).

[18]. This way of controlling sidelobes has the advantages of no signal loss due to weighting in frequency instead of amplitude. This nonlinear rate of change of frequency acts like amplitude weighting of the spectrum because it effectively reduces the spectrum energy near both ends of the waveform (shorter duration of each frequency near both ends due to high modulation rate). In addition, the mainlobe width only increases insignificantly. However the resulting system design is more complex than that of a LFM, and there is rare application of NLFM in civil radar systems.

It is worth noting that the sidelobe level of FM waveforms is not sensitive to the Doppler effects.

Phase Coded Waveforms

In phased coded waveforms the window function cannot be applied. One way to reduce the range sidelobes is through complementary code pairs as discussed in Section 3.4.1 for motion and gesture applications. Another category of methods is based on mismatched filter [19] concept and trying to find the optimal filters that provide minimal peak to sidelobe ratios. Various optimization techniques [20–25] have been explored to derive the optimal filter for different tradeoffs. There are also modified complementary code pairs based on mismatched filter concept to improve the Doppler tolerance [26–28]. It is worth noting that the Doppler effects have to be considered when developing sidelobe control techniques for phase coded waveforms. These researches are beyond the scope of this book, and the readers can refer to the corresponding references for further study.

4.2 Doppler Processing (Slow Time Processing)

For coherent radar, the data at each particular range gate can be processed across multiple pulses (phase coded waveform) or chirps (LFM waveform) to extract Doppler information after range processing. These waveforms form a coherent integration time (CIT) window, and spectrum analysis can be applied to these data inside. These waveforms are called pulse trains or chirp trains, as illustrated in Figure 4.7. Figure 4.7(a) shows the envelope of a chirp train within a CIT window; Figure 4.7(b) shows the envelope of a pulse train within a CIT window. The waveform repetition intervals (WRI) t_{wri} are labeled for chirp train and pulse train, respectively. The processing time or CIT is usually on the order of milliseconds, which is much longer than a single waveform. This is the reason that Doppler processing is referred to as slow time processing. The range data across multiple waveforms form a 2-D matrix as shown in Figure 4.8. At each range gate, the WRI between two consecutive waveforms are the sampling interval in slow time domain. After forming the 2-D data matrix, the signal processing is identical for LFM chirps and phase coded pulses.

As discussed in (3.1) in Chapter 3, coherent radar maintains consistent initial phase from chirp to chirp (or pulse to pulse). This means the return

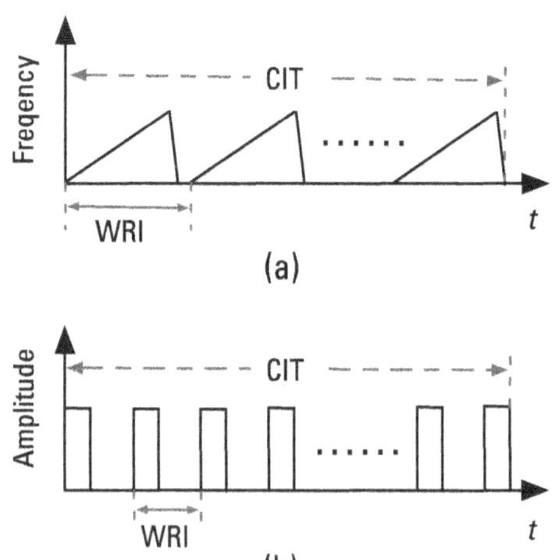

Figure 4.7 (a) Multiple chirp waveforms within a CIT; (b) multiple pulse waveforms within a CIT.

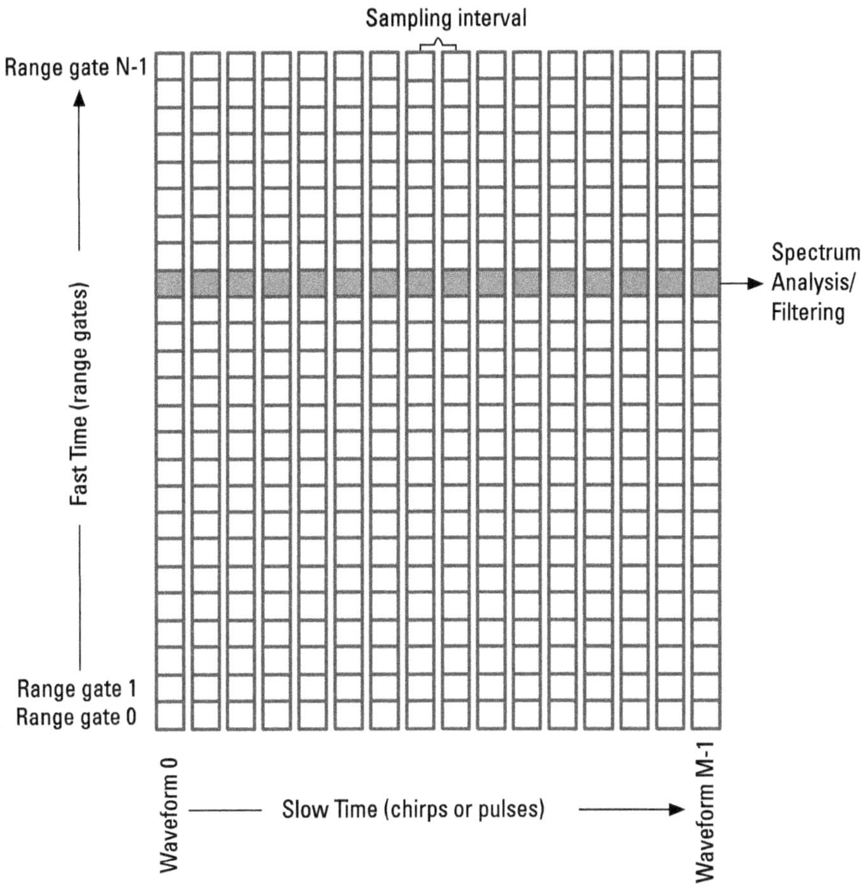

Figure 4.8 2-D data matrix of slow time data for each range gate.

signals from a stationary target will have the same phase across multiple waveforms, which forms the basis for Doppler processing. Let us assume there is a target at range gate n with a relative radial velocity v, and there are M waveform returns (samples) in the CIT window. The phase changes of data at range gate n between two consecutive waveforms are due to the range changes:

$$\Delta\theta = \frac{2\pi \bullet 2v \bullet t_{wri}}{\lambda} \tag{4.12}$$

where t_{wri} is the sampling interval as discussed earlier, and its inverse is the sampling frequency, factor 2 accounts for two-way range, and λ is the carrier wavelength.

Without loss of generality we can assume the first sample is a complex number of $Ae^{j\theta_0}$, and the mth sample can be express as:

$$s_r(n,m) = Ae^{j\theta_0}e^{j\frac{2\pi\bullet 2v\bullet t_{wri}\bullet m}{\lambda}} \quad 0 \le m \le M-1 \tag{4.13}$$

The discrete time Fourier transform (DTFT) [29] of the signal at the nth range gate is:

$$S_r(n,f) = \sum_{m=0}^{M-1} s_r(n,m)e^{-j2\pi fm\bullet t_{wri}} \tag{4.14}$$

Combining (4.13) and (4.14), we can obtain:

$$S_r(n,f) = Ae^{j\theta_0}\sum_{m=0}^{M-1} e^{j\frac{2\pi\bullet 2v\bullet t_{wri}\bullet m}{\lambda}} e^{-j2\pi fm\bullet t_{wri}} \tag{4.15}$$

where the summation term can be further simplified according to geometric sequence sum formula [30]. It should be noted that the DTFT is also the matched filter operation for Doppler processing of signal with form (4.13), and the result is optimal in terms of output SNR. The resulting DTFT spectrum can be expressed as:

$$S_r(n,f) = Ae^{j\theta_0}e^{-j2\pi\bullet t_{wri}\left(f-\frac{2v}{\lambda}\right)\frac{M-1}{2}} \frac{sin\left(\pi\bullet t_{wri}\left(f-\frac{2v}{\lambda}\right)M\right)}{sin\left(\pi\bullet t_{wri}\left(f-\frac{2v}{\lambda}\right)\right)} \tag{4.16}$$

Equation (4.16) indicates there is a peak (mainlobe) at frequency $f = \frac{2v}{\lambda}$ with amplitude of $M \bullet A$ according to L'Hospital's rule, and its location corresponds to the Doppler frequency of the target as defined in (3.37). Equation (4.16) is also called aliased sinc [31] function, and there are sidelobes around mainlobe and they all repeat at a period of $\frac{1}{t_{wri}}$ Hz in frequency domain (this is why it is also called periodic sinc function). This can be better explained through reformulating (4.13) as a product of a rectangular function, exponential function, and Dirac comb function:

$$s_r(n,m) = \left(rect\left(\frac{t-M\bullet t_{wri}/2}{M\bullet t_{wri}}\right)Ae^{j\theta_0}e^{j\frac{2\pi\bullet 2v\bullet t}{\lambda}}\right)\bullet\left(\sum_{m=-\infty}^{\infty}\delta(t-m\bullet t_{wri})\right) \tag{4.17}$$

where $\delta(t)$ is the Dirac delta function, and $\sum_{m=-\infty}^{\infty}\delta(t-m\bullet t_{wri})$ is the Dirac comb function. Multiplying any continuous signal by a Dirac comb function is often used to represent sampling process to obtain discrete time signal. The Fourier transform (FT) of the Dirac comb function is [29]:

$$\mathcal{F}\left(\sum_{m=-\infty}^{\infty}\delta(t-m\bullet t_{wri})\right)=\frac{1}{t_{wri}}\sum_{k=-\infty}^{\infty}\delta(f-k/t_{wri}) \tag{4.18}$$

According to [32], the FT of the product of two signals in time is the convolution of their FTs, and the FT of the first part of (4.17) is:

$$\begin{aligned}X(f)&=\mathcal{F}\left(rect\left(\frac{t-M\bullet t_{wri}/2}{M\bullet t_{wri}}\right)Ae^{j\theta_0}e^{j\frac{2\pi\bullet 2v\bullet t}{\lambda}}\right)\\&=\mathcal{F}\left(rect\left(\frac{t-M\bullet t_{wri}/2}{M\bullet t_{wri}}\right)\right)*\mathcal{F}\left(Ae^{j\theta_0}e^{j\frac{2\pi\bullet 2v\bullet t}{\lambda}}\right)\end{aligned} \tag{4.19}$$

where * indicates convolution. It is well known that the FT of a rectangular function is a sinc function, and the FT of an exponential function is a Dirac delta function [32]:

$$\mathcal{F}\left(rect\left(\frac{t-M\bullet t_{wri}/2}{M\bullet t_{wri}}\right)\right)=M\bullet t_{wri}\bullet e^{-j2\pi fM\bullet t_{wri}/2}sinc(\pi fM\bullet t_{wri}) \tag{4.20}$$

$$\mathcal{F}\left(Ae^{j\theta_0}e^{j\frac{2\pi\bullet 2v\bullet t}{\lambda}}\right)=Ae^{j\theta_0}\delta\left(f-\frac{2v}{\lambda}\right) \tag{4.21}$$

Combining (4.19) to (4.21) and considering the sifting property of Dirac delta function, the following holds:

$$X(f)=Ae^{j\theta_0}M\bullet t_{wri}\bullet e^{-j2\pi\left(f-\frac{2v}{\lambda}\right)M\bullet t_{wri}/2}sinc\left(\pi\left(f-\frac{2v}{\lambda}\right)M\bullet t_{wri}\right) \tag{4.22}$$

where $M\bullet t_{wri}$ is the overall coherent integration time T_{CIT}. The amplitude spectrum of (4.22) is a sinc function with its peak shifted to $\frac{2v}{\lambda}$ and the sidelobe level proportional to $1/f$.

The overall spectrum of (4.17) is the convolution of (4.18) and (4.22), and it can be obtained by repeating $\frac{1}{t_{wri}} \bullet X(f)$ along the f axis at intervals of $\frac{1}{t_{wri}}$ Hz and summing the results.

In summary, Doppler processing is to apply spectrum analysis along multiple waveform returns at each range gate, and the resulting spectrum is the so-called Doppler spectrum, which indicates how the echo power is distributed with respect to the targets' Doppler-shift frequencies. These waveform returns at each range gate can be considered as a sampled time sequence/signal, and discrete time signal processing techniques or principles can be directly applied. In practice, FFT is widely adopted for its efficiency and suitability for computing devices. Other high-resolution methods [33] such as ARMA, MUSIC, or maximum likelihood can also be applied if fine features of Doppler spectrum are of importance. Time-frequency analysis methods [34, 35] such as short-time Fourier transform (STFT) and wavelet transform (WT) have also been utilized to extract micro-Doppler features.

4.2.1 Sampling Frequency in Slow Time Domain

The Nyquist sampling theory should be used to guide the choice of sampling interval or WRI in that the return data from multiple waveforms at each specific range gate can be considered as being obtained through sampling a Doppler signal in slow time domain. This sampling interval is generally constant within a CIT window resulting in uniform samples in time domain. Nonuniform sampling is rarely used due to its associated complex processing and high sidelobes.

The corresponding sampling frequency or waveform repetition frequency (WRF) $f_{wrf} = 1/t_{wri}$ needs to be at least twice of the maximum Doppler frequency of interests to avoid spectrum aliasing. Each application will have its own specific requirements on the maximum Doppler frequency of interests. Figure 4.9 shows typical Doppler zones of various civil activities. For human motion detection, the maximum velocity of interest is about 2.5 m/s, and for gesture recognition the maximum velocity of interest is about 4.5 m/s [36]. For

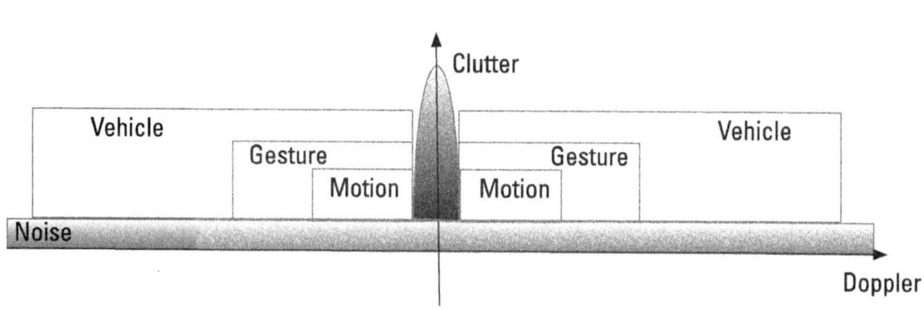

Figure 4.9 Doppler zones of typical activities.

automotive radar, the maximum velocity can be more than 70 m/s when two cars are heading toward each other on a divided high way. For 60-GHz motion sensing radar, the WRF needs to be at least 2,000 Hz for motion sensing; for 76-GHz far range automotive radar, the WRF needs to be at least 71.1 KHz, according to (3.37) and Nyquist theory.

The upper bound of the WRF is generally limited by hardware limitation or the maximum unambiguous range beyond which targets are expected. The maximum unambiguous range is determined by:

$$R_u = t_{wri} \bullet \frac{c}{2} = \frac{c}{2 \bullet f_{wrf}} \tag{4.23}$$

If the unambiguous range R_u is too small, there might be echoes reflected from large targets or clutter beyond R_u being received in the next waveform's listening time. These echoes are often called second time around returns and illustrated in Figure 4.10. Target B is beyond the unambiguous range and will not be received during the first waveform listening period. Its return is folded to the apparent range, which is a modulo function of true range during the second

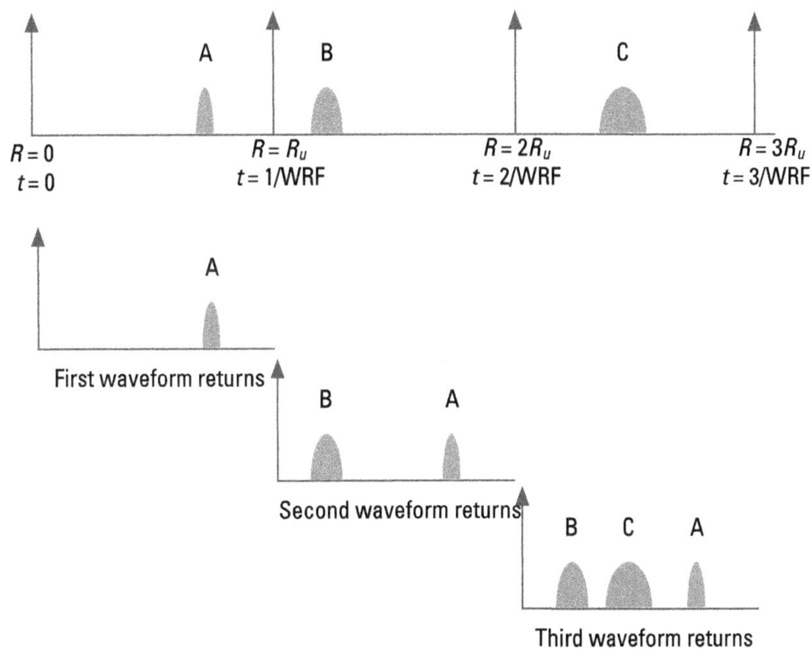

Figure 4.10 Multiple time around returns that can be mistaken as a short-range echo of the next receiving cycles. Target A is within the unambiguous range, target B is a second time around return, and target C is a multiple (third) time around return. On the third waveform listening period, targets B and C show up at ambiguous close in ranges.

waveform listening period. Target C is an example of a multiple time around returns, which will be received during the third waveform listening period.

The second/multiple time around returns are less concerning in civil radar applications. The range of interests is generally less than 10% of the unambiguous range. Compared to these target returns within the range of interests, the echoes from targets beyond unambiguous range are subject to significant range attenuation, which is inversely proportional to range to the fourth power (>40 dB based on radar range equation).

Waveform Duration Constraints

The requirement on WRF/WRI sets a hard limit on the maximum duration of waveforms. For 2,000-Hz WRF, the maximum length of each individual waveform needs to be less than 0.5 ms. For 71.1-kHz WRF, the maximum length of each individual waveform needs to be less than 14.1 μs. The minimum duration of waveforms is determined by link budget requirements, as discussed in Chapter 2.

WRF Staggering

The WRF can be different from CIT to CIT, and this is referred to as WRF staggering. It is useful to resolve range ambiguities due to multiple around time returns and/or Doppler ambiguity due to high radial velocity target returns.

According to (4.23), different WRFs correspond to different maximum unambiguous ranges. The returns from targets within the unambiguous range (such as Target A in Figure 4.10) remain at their true range regardless of the WRF changes. The returns from targets beyond the unambiguous range appear at different apparent range for each WRF. Theoretically the true range can be resolved with a pair of WRFs. Let us assume the first CIT adopts $f_{wrf} = f_1$ with an unambiguous range of R_{u1}, and the measured apparent range of the multiple around time target is at R_1. Then the true range of this target is determined by one of the k values in the following equation:

$$R_{T1} = R_1 + k \bullet R_{u1},\ \ k = 0,1,2\ldots k_{max1} \tag{4.24}$$

where k_{max1} is an integer determined by the ratio of the maximum instrumented range (range of interest) to the maximum unambiguous range k_{max1} = round($R_{\max}/R_{u1}$). If the second CIT uses $f_{wrf} = f_2$ with an unambiguous range of R_{u2}, and the measured apparent range of the multiple around time target is at R_2, then the true range of this target is determined by one of the k values in the following equation:

$$R_{T1} = R_2 + k \bullet R_{u2},\ \ k = 0,1,2\ldots k_{max2} \tag{4.25}$$

where $k_{max2} = \text{round}(R_{max}/R_{u2})$. The true range can be determined through solving (4.24) and (4.25) for a pair of k values, whose corresponding true ranges are within an error window of two or three times of the range measurement accuracy. In practice a third WRF is often used to improve the accuracy and reduce the false values [37]. The Chinese Remainder Theorem (CRT) [38] is a useful tool to find the unambiguous range. Let us use the number of range bins and assume there are L WRFs and the corresponding maximum unambiguous range bins r_{ui} are pairwise prime integers. Under such circumstances, CRT provides a way to directly calculate the true range bin expressed by the congruence [39]. The true range bins r_T and measured range bins r_i have the following congruent relationship:

$$r_T \equiv r_1 \left(mod\ r_{u1}\right), r_T \equiv r_2 \left(mod\ r_{u2}\right), r_T \equiv r_L \left(mod\ r_{uL}\right) \tag{4.26}$$

Then r_T is the unique remainder after division by $r_u = r_{u1} \bullet r_{u2} \ldots r_{uL}$:

$$r_T = \left(\sum_{i=1}^{L} C_i r_i\right)\left(modulo\ r_u\right) \tag{4.27}$$

where

$$C_i = b_i r_u / r_{ui} \tag{4.28}$$

and b_i is the smallest integer meeting the following condition:

$$b_i \frac{r_u}{r_{ui}} \left(mod\ r_{ui}\right) = 1 \tag{4.29}$$

For example, if there are three WRFs and the corresponding maximum unambiguous range bins are $r_{u1} = 5$, $r_{u2} = 7$, and $r_{u3} = 9$. According to (4.29), $b_1 = 2$, $b_2 = 5$, and $b_3 = 8$. According to (4.28), $C_1 = 126$, $C_2 = 225$, and $C_3 = 280$. If the measurements are $r_1 = r_2 = r_3 = 3$, and there is no range ambiguity and (4.27) also reports the true range bin $r_T = 3$. If the measurements are $r_1 = 2$, $r_2 = 3$, and $r_3 = 4$, the true range bin is $r_T = 157$ according to (4.27).

Similar ideas can be applied to address the ambiguity in Doppler domain. For example, if two different WRFs, 2,000 Hz and 2,500 Hz, are chosen alternately from CIT to CIT and a target return with Doppler frequency of 3,000 Hz are in the radar field of view, the aliased (fold over) Doppler frequency of the target appears to be 1,000 Hz for the 2,000 Hz WRF and 500 Hz for the 2,500 Hz WRF, respectively. In a similar way to (4.24) and (4.25), the true velocity can be estimated. The selection of the WRFs requires careful considerations to minimize false velocity estimation. This technique has been widely adopted in air traffic control radars, and the design principles are well explained in [40, 41].

The WRF staggering method to resolve range or Doppler ambiguities requires multiple detections from the same target over a few CITs (minimum two). The CITs must also use different WRFs. In practice, it is often to utilize more CITs with different WRFs and select the best a few to process in order to handle clutter, blind velocity, and so on.

4.2.2 CIT Window Size

It will be discussed later in this chapter that the CIT window length determines the resolution of the Doppler frequency. It is desirable to have fine Doppler frequency cells for gesture recognition. However, there are constraints to limit the maximum CIT window length: one is called range walking and the other one is coherence time.

Range walking refers to the situation where the target of interest moves more than one range resolution cell during the CIT window. It is obvious that the maximum integration time T_{CIT} is limited by the maximum radial velocity and the range resolution cell size:

$$T_{CIT} \leq R_{res} / v_{rmax} \tag{4.30}$$

where a rigid target has tangential motion with aspect angle changing, the coherence time can be estimated [42] by the following:

$$T_C = \frac{\lambda}{2\omega L} \tag{4.31}$$

where λ is wavelength, ω is the angle velocity, and L is the target dimension at across range. For the elastic targets, the correlation time can be determined through experiments with their Doppler spectrum:

$$T_C = \frac{1}{2\pi\sigma_D} \tag{4.32}$$

where σ_D is the standard deviation of the target's Doppler spectrum. T_{CIT} should be bounded by the coherence time defined by either (4.31) or (4.32).

4.2.3 MTI and Clutter Cancellation

Prior to Doppler processing, it is common to apply high pass filter to remove the DC leakage and clutter at low Doppler frequencies. This technique is referred by radar engineers as moving target indication (MTI). When these leakage and clutter are strong, MTI is often necessary even if Doppler processing will be applied at a later stage. The reason is that Doppler processing often suffers sidelobe issues, and the high Doppler targets may have to compete with low Doppler clutter through sidelobes. The MTI filter is applied in the slow (discrete) time domain, and its frequency response is periodic with period equal to the *WRF* as discussed before. This means that high-speed targets with Doppler frequencies equal to multiples of the WRF will be filtered out, too. This issue is known as the blind speed problem, which can be addressed through staggered WRFs [43]. The blind speed is determined by:

$$v_b = \frac{f_{wrf} \bullet \lambda}{2} \tag{4.33}$$

The MTI filters are generally finite impulse response (FIR) filters [44] and can have a minimum length of 2 and maximum length of the available data inside the CIT window. An example is shown in Figure 4.11 for an MTI filter of order two (two delay lines/three taps). The optimal coefficients are derived in [45] based on the criterion that maximizes the signal-to-clutter ratio at the output of the filter divided by the signal-to-clutter ratio at the input of the filter,

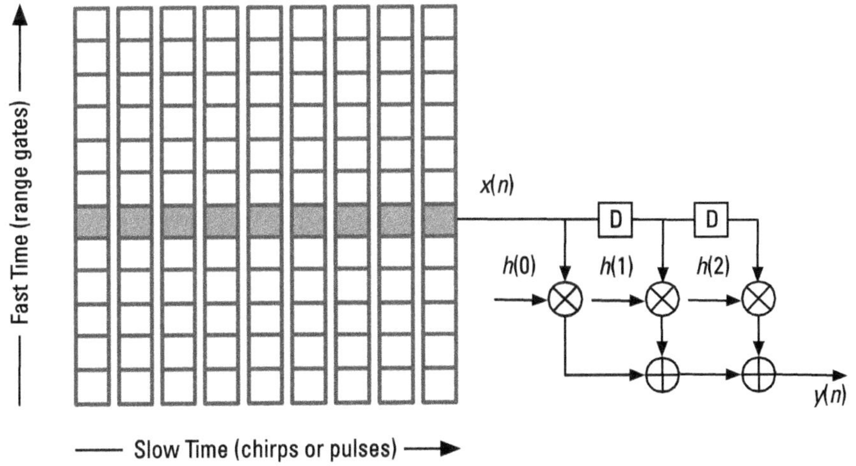

Figure 4.11 Illustrative example of a 2-delay line MTI filter.

averaged uniformly over all target radial velocities of interest. The averaged ratio of these two signal-to-clutter ratios is also known as improvement factor.

In practice, binomial coefficients of alternating signs are often applied instead of optimally designed filters in that they have nearly identical performance [46]. These coefficients can be expressed as:

$$h(n) = (-1)^m \frac{(K-1)!}{m!(K-m-1)!},\ m = 0,1,\ldots,K-1 \tag{4.34}$$

where K is the number of taps in the MTI filter, and $(K-1)!$ indicates factorial of (K – 1). The frequency responses of binomial coefficients based MTI filters have the following general form:

$$H(f) = \left(1 - e^{-j2\pi f/f_{wrf}}\right)^{K-1} = e^{-j\pi f(K-1)/f_{wrf}} (2j)^{K-1} sin^{K-1}\left(\pi f / f_{wrf}\right) \tag{4.35}$$

and the magnitude frequency response:

$$\left|H(f)\right| = 2^{K-1}\left|sin^{K-1}\left(\pi f / f_{wrf}\right)\right| \tag{4.36}$$

The magnitude frequency response for 2, 3, and 5 taps MTI filters are plotted in Figure 4.12, where these responses are normalized to the average power gain for noise. It is obvious that the higher the filter order, the wider the null width. It is important to use higher order MTI filter when there is strong low motion clutter or smearing of stationary clutter due to antenna motion. The tradeoff is the possible suppression of low-speed target and loss of samples for the following Doppler processing. For example, if there are 32 waveforms in one CIT, then there will be only 28 samples after a 5 tap MTI filter. There is also noise correlation loss, which is due to the reduction of independent noise samples at the output of the filter. This loss results in the increase of the detectability factor or effectively the reduction of SNR. A simulation study [47] has shown that the noise correlation loss ranges from about 1 dB for 2 tap MTI filter to 2.5 dB for 5 tap MTI filter at probability of detection of 0.9 and false alarm rate of 10^{-6}.

It is worth noting that infinite impulse response (IIR) filters or recursive filters can achieve better clutter rejection with fewer delay lines. However, stability concerns need to be addressed when applying IIR filters to real radar systems in that they can have poor transient response due to the recursive nature. Large interference or clutter can produce long lasting transient ringing and mask real targets.

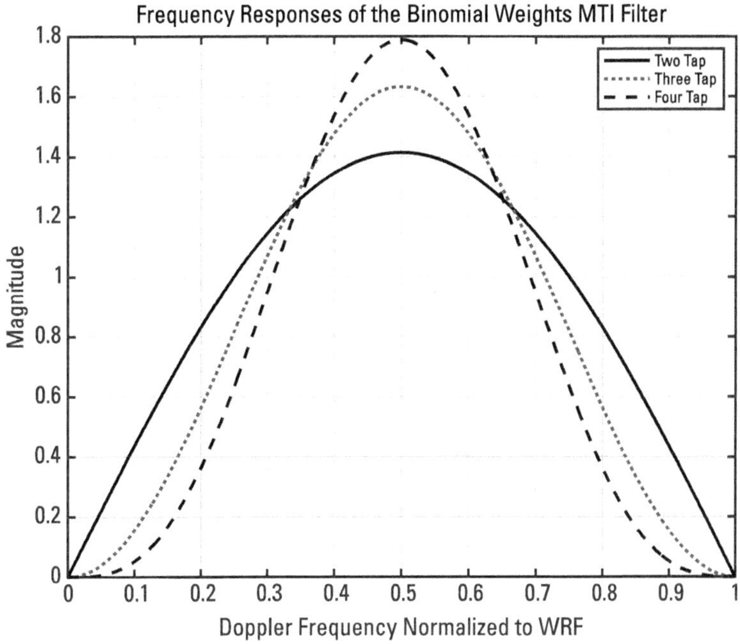

Figure 4.12 Doppler frequency response of MTI filters (cancelers) normalized to noise gain and WRF.

Stationary clutter and DC leakage filtering can also be achieved through estimating these signals with a recursive first order filter:

$$C(n) = \alpha C(n-1) + (1-\alpha)x(n-1) \tag{4.37}$$

and the clutter cancellation filtering is:

$$y(n) = x(n) - C(n) \tag{4.38}$$

where $C(n)$ is the estimate of the clutter at the nth waveform, $x(n)$ is the data output at the nth waveform, and $y(n)$ is the filtered data with clutter being removed. This way of estimating the clutter $C(n)$ can be considered as a clutter map of zero Doppler frequency filter and is updated continuously across CIT boundaries. $\alpha < 1$ is the scale factor that determines the time constant of estimating $C(n)$. The physical meaning of α can be explained from the analog counterpart of the low pass filter (4.37). The corresponding first order analog filter has the following continuous transfer function in Laplace transform domain:

$$H(s)=\frac{1}{1+s\tau}=\frac{1/\tau}{s+1/\tau} \tag{4.39}$$

where τ is the time constant and used to indicate how rapidly an exponential function decays. Step invariant method can be used to transform the previous analog filter into discrete time filter whose step response is the sampled version of the corresponding analog filter. The general equation of step invariant method [48] is as follows:

$$\hat{H}(z)=\frac{z-1}{z}\mathcal{Z}\left[\mathcal{L}^{-1}\left[\frac{H(s)}{s}\right]\right] \tag{4.40}$$

where $\mathcal{Z}$ indicates Z-transform in discrete time domain, and $\mathcal{L}^{-1}$ represents inverse Laplace transform. The Laplace transform of step function $u(t)$ is:

$$\mathcal{L}[u(t)]=1/s \tag{4.41}$$

and the Z-transform of step function $u(n)$ is:

$$\mathcal{Z}[u(n)]=\frac{z}{z-1} \tag{4.42}$$

Considering (4.41) and (4.42), (4.40) can be rearranged and interpreted as:

$$\hat{H}(z)\mathcal{Z}[u(n)]=\mathcal{Z}\left[\mathcal{L}^{-1}\left[H(s)\mathcal{L}[u(t)]\right]\right] \tag{4.43}$$

where $\hat{H}(z)\mathcal{Z}[u(n)]$ is the step response in discrete time domain and $H(s)\mathcal{L}[u(t)]$ is the step response in continuous time domain.

With step invariance method and Laplace-Z transform table [49], the analog filter (4.39) has the following discrete time domain counterpart:

$$\hat{H}(z)=\frac{z-1}{z}\mathcal{Z}\left[\mathcal{L}^{-1}\left[\frac{1/\tau}{s(s+1/\tau)}\right]\right]=\frac{z-1}{z}\frac{z\left(1-e^{-T/\tau}\right)}{(z-1)\left(z-e^{-T/\tau}\right)}=\frac{1-e^{-T/\tau}}{z-e^{-T/\tau}} \tag{4.44}$$

where T is the sampling period for the discrete time domain, and $T=t_{wri}$ for our application.

The transfer function of (4.37) is:

$$H(z)=\frac{(1-\alpha)z^{-1}}{1-\alpha z^{-1}}=\frac{1-\alpha}{z-\alpha} \tag{4.45}$$

Comparing (4.44) and (4.45), it is obvious that the following relationship holds:

$$\alpha=e^{-T/\tau} \tag{4.46}$$

Equation (4.46) provides guidance on how to set α. For example, if we want to effectively use 32 waveforms to estimate the current clutter value, then α should be set as:

$$\alpha=e^{-T/32T}=e^{-1/32}=0.97 \tag{4.47}$$

It is straightforward to implement (4.37) and reduce the memory requirements significantly comparing to simple running average. As for the example of (4.47), only one value needs to be stored in the memory instead of 31.

Combining (4.37) and (4.38) will lead to the overall filter:

$$y(n)-\alpha y(n-1)=x(n)-x(n-1) \tag{4.48}$$

whose frequency response is:

$$h\left(e^{jwn}\right)=\frac{1-e^{-jwn}}{1-\alpha e^{-jwn}} \tag{4.49}$$

The discussion of MTI is based on a stationary platform where the radar system itself is not moving. When the radar is on a moving platform, the Doppler frequency of the clutter is determined by the relative velocity of the platform and the clutter. The Doppler spreading due to these motions can also degrade the MTI filter performance. Various techniques have been developed to deal with the nonzero Doppler shift and spectrum spreading before applying the MTI filter. The readers can refer to [50] for clutter Doppler shift compensation, [51] for Doppler spectrum spread reduction, and [52] for a more effective STAP method, which filters the clutter in both spatial and temporal domains jointly.

4.2.4 Moving Target Detector (Filter Bank)

When there are a number of samples available in each CIT, the filter bank is generally implemented to provide better detectability especially in nonstationary clutter situations. This technique is referred to as moving target detector (MTD). MTD was first developed [53] for air traffic control radars, which has improved moving targets detection in both stationary ground clutter and nonstationary weather clutter. MTD architecture is widely adopted in the civil applications such as automotive radars and motion detection and gesture control radars. In these civil applications, there are many waveforms transmitted and received within each CIT; the number of filters is large in the filter bank. Under such circumstances, discrete Fourier transform (DFT) is often adopted to be the desired filter bank. Each point of the DFT output is corresponding to a filter output in the filter bank. M point DFT is equivalent to a filter bank with M filters. As a common practice, FFT with windowing is adopted to reduce computation complexity. The example shown in Figure 4.8 has M waveforms transmitted in each CIT, and after passing through a filter bank with M filters the results are a 2-D spectrum matrix, as shown in Figure 4.13. Note the number of filters in the filter bank can be any other number, although it is usually the same as the number of waveforms in the CIT when FFT is applied as the filter bank.

When the number of waveforms is limited within a CIT, specially designed filter banks are often used to control the various clutter situation as those in recent air traffic control radars [54, 55].

4.2.5 Doppler (Radial Velocity) Resolution

Doppler resolution is the capability to resolve two targets that have the same range but different Doppler frequencies. It is the required minimum radial speed difference with which two equal RCS scatterers can be separated and discerned as individual scatterers. Doppler resolution is also associated with a likelihood to indicate how likely the two scatterers can be resolved due to the stochastic nature of radar signals. Similar to range resolution, the Doppler

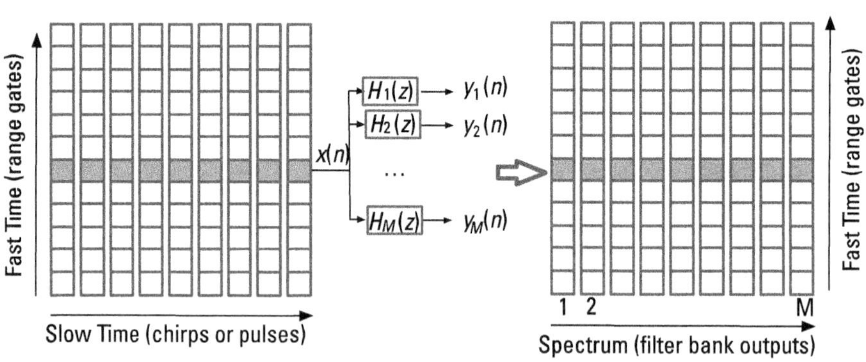

Figure 4.13 Illustration of range-Doppler spectrum matrix formation at the output of a filter bank.

resolution is generally worse when the two scatterers have different RCS. Strict mathematical treatment of the Doppler resolution is not trivial, and the readers can refer to [56] for details. In the following discussion, we will assume that the scatterers reflect equal amplitude return signals that are strong enough to be detected.

When FFT is applied, the common definition of Doppler resolution is based on the overall integration time, which is consistent with the well-known Fourier analysis theory:

$$D_{res} = 1 / T_{CIT} \tag{4.50}$$

and the corresponding radial velocity resolution is:

$$V_{res} = \lambda / (2T_{CIT}) \tag{4.51}$$

where λ is the wavelength of the carrier. When Doppler/radial velocity resolution is referenced in radar systems specifications, it is generally referring to (4.50) and (4.51). This resolution definition can also be considered as based on Rayleigh criterion in that the Doppler spectrum after DFT takes the form of sinc function with the first zero at $1/T_{CIT}$ according to (4.22). The associated probability of resolution is ~0.7, which is different from that of the range resolution examples discussed in Section 4.1.3, where the matched filter output is assumed to have a triangle shape instead of a sinc shape.

When the signal-to-noise ratio is large and the two target returns are not coherent, other high-resolution methods (such as subspace based methods and maximum likelihood method) can be adopted to significantly improve the Doppler resolution at the cost of computation complexity. However these high-resolution methods mainly find applications to improve angle resolution instead of Doppler resolution in that the DFT-based method provides good enough resolution in practice and is much simpler to implement.

4.2.6 Doppler (Radial Velocity) Accuracy

The Doppler accuracy δD is derived based on inverse probability in [57, 58] and summarized in [11] as the following:

$$\delta D = \frac{1}{\alpha\left(2E / N_0\right)^{1/2}} \tag{4.52}$$

where E is the signal energy, N_0 is the noise power per unit bandwidth, and α is called the effective time duration of the signal and defined as:

$$\alpha^2 = \frac{1}{E}\int_{-\infty}^{\infty} (2\pi t)^2 s^2(t)\, dt \tag{4.53}$$

where $s(t)$ is the input signal in time. Doppler accuracy has similar expressions to that of range accuracy in (4.7) and (4.8). The corresponding radial velocity accuracy is:

$$\delta V = \frac{\lambda \delta D}{2} \tag{4.54}$$

It is obvious that the Doppler accuracy is inversely proportional to the effective time duration and square root of signal-to-noise ratio. The value of α is determined by the shape of the signal, and various examples are shown in [11], and for a simple rectangular pulse the Doppler accuracy is:

$$\delta D = \frac{\sqrt{3}}{\pi\tau\left(2E / N_0\right)^{1/2}} \tag{4.55}$$

where τ is the pulse length. When there are M such pulses used in a CIT, the corresponding accuracy is:

$$\delta D = \frac{\sqrt{3}}{\pi T_{CIT}\left(2E / N_0\right)^{1/2}} = \frac{\sqrt{3}D_{res}}{\pi\left(2E / N_0\right)^{1/2}} \tag{4.56}$$

It is clear that the Doppler accuracy depends on both Doppler resolution and the signal-to-noise ratio. As a rule of thumb, it is often safe to assume a 10% of Doppler resolution as the Doppler accuracy, which corresponds to 12-dB signal-to-noise ratio target.

4.2.7 Doppler Sidelobes Control

When DFT is applied, there are sidelobes of –13.2 dB with respect to the main lobe in the corresponding Doppler spectrum. This phenomenon has been well explained in many Fourier theory references. The sidelobes can be a problem when strong targets and weak targets coexist at the same range cell. The returns from weak target may be masked by the sidelobes of strong targets even when they have different radial velocities. This is the reason window function is almost always being applied prior to Doppler processing in practice. Similar to the discussion in Section 4.1.5 on FM waveform time sidelobe control, this problem is also analogous to spectral leakage in Fourier analysis of a function with finite length. The windows such as Hamming, Blackman, Dolph-Chebyshev, Taylor, and many others [14] used for spectrum leakage control are directly applicable here. The main tradeoffs are broader mainlobe and signal loss in the processing gain versus the sidelobe levels. The signal loss will affect the final signal-to-noise ratio and detection performance, and the mainlobe increase means degraded Doppler resolution and accuracy. The readers can refer to Section 4.1.5 on how to select the window functions.

4.3 Summary

Range processing separates target returns into different range gates. At each range gate, the Doppler processing integrates multiple waveforms to achieve integration gain and improve SNR. It also separates targets from clutter based on different radial velocities. After processing, targets can be detected and resolved if they are at either different range or Doppler from each other and clutter.

DFT is the most popular and important filter bank in civil radar applications. With proper window function, the resultant Doppler spectrum has a good tradeoff among coherent integration gain, sidelobes, clutter separation, and computational efficiency.

The range and Doppler accuracies correlate closely with their corresponding resolutions. Improving resolutions will also improve the corresponding accuracy. However, accuracy can also be improved through maximizing the SNR, which is not the case for resolution.

References

[1] Skolnik, M. I., *Introduction to Radar Systems*, Third Edition, Chapter 6.5, New York: McGraw-Hill, 2001.

[2] Rihaczek, W., *Principles of High-Resolution Radar*, Peninsula Publishing, 1985.

[3] Ducoff, M. R., "Pulse Compression Radar," in *Radar Handbook*, Third Edition, Chapter 8, New York: McGraw-Hill, 2008.

[4] Skolnik, M. I., *Introduction to Radar Systems*, Third Edition, Chapter 5.7, New York: McGraw-Hill, 2001.

[5] Born, M., and E. Wolf, *Principles of Optics*, Cambridge, UK: Cambridge University Press, 1999, p. 4.61.

[6] Trunk, G. V., "Range Resolution of Targets," *IEEE Trans. AES-20*, November 1984, pp. 789–797.

[7] Trunk, G. V., "Range Resolution of Targets Using Automatic Detectors," *IEEE Trans. AES-14*, September 1978, pp. 750–755.

[8] Curry, G. R., *Radar System Performance Modeling*, Second Edition, Chapter 8.1, Norwood, MA: Artech House, 2005.

[9] Woodward, P. M., *Probability and Information Theory, with Applications to Radar*, New York: McGraw-Hill, 1953.

[10] Weiss, A. J., "Composite Bound on Arrival Time Estimation Errors," *IEEE Transactions on Aerospace and Electronic Systems*, Vol. AES-22, No. 6, Nov. 1986, pp. 751–756.

[11] Skolnik, M. I., *Introduction to Radar Systems*, Third Edition, Chapter 6.3, New York: McGraw-Hill, 2001.

[12] Oppenheim, A. V., et al., *Signals and Systems*, Second Edition, Pearson Education,1996.

[13] Levanon, N., et al., *Radar Signals*, Chapter 4, New York: John Wiley & Sons, 2004.

[14] Harris, F. J., "On the Use of Windows for Harmonic Analysis with the Discrete Fourier Transform," *Proceedings of the IEEE*, Vol. 66, No. 1, 1978, pp. 51–83.

[15] Peeples, P., *Radar Principles*, Chapter 7.3, New York: John Wiley & Sons, 2004.

[16] Lyons, R. G., *Understanding Digital Signal Processing*, Second Edition, Chapter 13, Upper Saddle River, NJ: Prentice Hall, 2004.

[17] Barton, D. K., *Radar System Analysis and Modeling*, Chapter 5, Norwood, MA: Artech House, 2005.

[18] Wang, J., E. Brookner, et al., "Modernization of En Route Air Surveillance Radar," *IEEE Transactions on Aerospace and Electronic Systems*, Vol. 48, No. 1, Jan. 2012, pp. 103–115.

[19] Ackroyd, M. H, and F. Ghani, "Optimum Mismatched Filters for Sidelobe Suppression," *IEEE Transactions on Aerospace and Electronic Systems*, Vol. AES-9, No. 2, 1972, pp. 214–218.

[20] Baden, J., Cohen, M., "Optimal Peak Sidelobe Filters for Biphase Pulse Compression," *Proceedings of the IEEE International Radar Conference*, 1990, pp. 249–252.

[21] Griep, K., J. Ritcey, and J. Burlingame, "Poly-Phase Codes and Optimal Filters for Multiple User Ranging," *IEEE Transactions on Aerospace and Electronic Systems*, Vol. 31, No. 2, 1995, pp. 752–767.

[22] Cilliers, J., and J. Smit, "Pulse Compression Sidelobe Reduction by Minimization of Lp-Norms," *IEEE Transactions on Aerospace and Electronic Systems*, Vol. 43, No. 3, 2007, pp. 1238–1247.

[23] De Maio, A., et al., "Design of Radar Receive Filters Optimized According to lp-Norm Based Criteria," *IEEE Transactions on Signal Processing* , Vol. 59, No. 8, 2011, pp. 4023–4029.

[24] Stoica, P., J. Li, and M. Wue, "Transmit Codes and Receive Filters for Radar," *IEEE Signal Processing Magazine*, Vol. 25, No. 6, 2008, pp. 94–109.

[25] Rabaste, O., Savy, L., "Mismatched Filter Optimization for Radar Applications Using Quadratically Constrained Quadratic Programs," *IEEE Transactions on Aerospace and Electronic Systems*, Vol. 51, No. 4, 2015, pp. 3107–3122.

[26] Kretschmer, F. F., and B. L. Lewis, "Doppler Properties of Polyphase Coded Pulse-Compression Waveforms," *IEEE Transactions on Aerospace and Electronic Systems*, Vol. AES-19, No. 4, 1983, pp. 521–531.

[27] Pezeshki, A., et al., "Doppler Resilient Golay Complementary Waveforms," *IEEE Transactions on Information Theory*, Vol. 54, No. 9, 2008, pp. 4254–4266.

[28] Nguyen, H. D., and G. . Coxson, "Doppler Tolerance, Complementary Code Sets, and Generalised Thue–Morse Sequences," *IET Radar, Sonar & Navigation*, Vol. 10, No. 9, 2016, pp. 1603–1610.

[29] Oppenheim, A. V., and R. W. Schafer, *Discrete-Time Signal Processing*, Second Edition, Upper Saddle River, NJ: Prentice Hall, pp. 60, 1999.

[30] Ivanova, O. A., "Geometric Progression," Encyclopedia of Mathematics, http://encyclopediaofmath.org/index.php?title=Geometric_progression&oldid=12512 (last accessed May 2022).

[31] Burrus, C. S., et al., *Computer-Based Exercises for Signal Processing Using MATLAB*, Upper Saddle River, NJ: Prentice Hall, 1994.

[32] Kammler, D., *A First Course in Fourier Analysis*, Upper Saddle River, NJ: Prentice Hall, 2000.

[33] Scharf, L. L., *Statistical Signal Processing—Detection, Estimation, and Time Series Analysis*, Addison-Wesley Publishing Company, 1991.

[34] Chui, C., *An Introduction to Wavelets*, First Edition, Academic Press, 1992.

[35] Cohen, L., *Time–Frequency Analysis*, New York: Prentice-Hall, 1995.

[36] Elgendi, M., et al., "Real-Time Speed Detection of Hand Gesture Using Kinect," The 25th Annual Conference on Computer Animation and Social Agents, Singapore, 2012.

[37] Skolnik, M. I., *Introduction to Radar Systems*, Third Edition, Chapter 2.10, New York: McGraw-Hill, 2001.

[38] Dence, J. B., and T. P. Dence, *Elements of the Theory of Numbers*, Chapter 4, Academic Press, 1999.

[39] Trunk, G., and S. Brockett, "Range and Velocity Ambiguity Resolution," *IEEE National Radar Conference*, 1993, pp. 146–149.

[40] Skolnik, M. I., *Introduction to radar Systems*, Third Edition, Chapters 3.3 and 3.6, New York: McGraw-Hill, 2001.

[41] Shrader, W. W., "MTI Radar," in *Radar Handbook*, Third Edition, Chapter 2, New York: McGraw-Hill, 2008.

[42] Barton, D. K., *Radar System Analysis and Modeling*, Chapter 2.5, Norwood, MA: Artech House, 2005.

[43] Barton, D. K., *Radar System Analysis and Modeling*, Chapter 5.3, Norwood, MA: Artech House, 2005.

[44] Oppenheim, A. V., and R. W. Schafer, *Discrete-Time Signal Processing*, Chapter 6.5, Second Edition, Upper Saddle River, NJ: Prentice Hall, 1999.

[45] Capon, J., "Optimum Weighting Functions for the Detection of Sampled Signals in Noise," *IEEE Transactions on Information Theory*, Vol. 10, No. 2, 1964, pp. 152–159.

[46] Andrews, G. A., "Optimal Radar Doppler Processors," *NRL Report 7727*, Washington, D.C., May 1974.

[47] Trunk, G. V., "MTI Noise Integration Loss," *Proceedings of the IEEE*, Vol. 65, No. 11, Nov. 1977, pp. 1620–1621.

[48] Morin, A., and P. Labbe, "Derivation of Recursive Digital Filters by the Step-Invariant and the Ramp-Invariant Transformations," *DREV Report 4325/84*, DRDC, May 1984.

[49] Isermann, R., *Digital Control Systems*, Volume 2, Appendix A: Table of Laplace and z-Transforms, Berlin Heidelberg: Springer-Verlag, 1991.

[50] Day, J. K., et al., "Airborne MTI," in *Radar Handbook*, Third Edition, Chapter 3.4, New York: McGraw-Hill, 2008.

[51] Day, J. K., et al., "Airborne MTI," in *Radar Handbook*, Third Edition, Chapter 3.5, New York: McGraw-Hill, 2008.

[52] Guerci, J. R., *Space-Time Adaptive Processing for Radar*, Second Edition, Norwood, MA: Artech House, 2014.

[53] Muehe, C. E., "Digital Signal Processor for Air Traffic Control Radars," *IEEE NEREM 74 Record, Part 4: Radar Systems and Components*, October 1974, pp. 73–82.

[54] Sergey, L., et al., "Advanced Mitigating Techniques to Remove the Effects of Wind Turbines and Wind Farms on Primary Surveillance Radars," *2008 IEEE Radar Conference*, Rome, Italy, 2008, pp. 1–6.

[55] Wang, J., et al., "Modernization of En Route Air Surveillance Radar," *IEEE Transactions on Aerospace and Electronic Systems*, Vol. 48, No. 1, 2012, pp. 103–115.

[56] Peeples, P., *Radar Principles*, Chapter 8.2, New York: John Wiley & Sons, 2004.

[57] Manasse, R., "Range and Velocity Accuracy from Radar Measurements," Lincoln Laboratory Memo, February 1955.

[58] Manasse, R., "Parameter Estimation Theory and Some Applications of the Theory to Radar Measurements," *Mitre Technical Series Report No. 3*, 1960.

5

Array Signal Processing

Many radar systems have multiple transmit and/or receive antennas. Array signal processing takes advantage of the spatial information collected from these antennas to extract angle information of targets. The data from these antennas can also be utilized to enhance the signal-to-noise ratio of targets and improve detectability.

The antenna array can have various structures. The simplest form is linear (line) array, where all antenna elements are arranged in a straight line to provide azimuth capability. The linear array can be extended to 2-D planar array to provide both azimuth and elevation information. The shape of the planar array can be rectangle, circular, or even hexagonal. There is also conformal array where the antennas are arranged on curved surface, which is not common in practice due to manufacturing and calibration challenges. In this chapter, we will discuss the most popular linear array and rectangular planar array and particularly focus on uniform arrays where the elements and their separations are identical. These array structures are simpler to implement and possess efficient data processing algorithms. They meet most applications' requirements and are robust to real-world challenges and limitations.

There have been extensive research and applications of array processing techniques in the past few decades [1–6]. Thorough descriptions will probably require a whole book; therefore, in this chapter we will limit our discussions to those techniques applied in real civil radar systems or having high potential to be fielded in the near future. Among these techniques, beamforming is the conventional and widely adopted one, which directs the majority of RF energy to a particular direction to improve the signal-to-noise ratio and determine the direction of targets. Beamforming can be applied to transmit and/or to receive in analog or digital ways. Nowadays digital beamforming on receive is a standard configuration in civil radar applications due to the availability of low-cost receiving subsystems. Beamforming is analogous to Fourier analysis and thereafter has the same limitation on resolution. There are many complex algorithms to improve the angle resolution based on accurately modeling the received signals, and these methods are referred to as high-resolution methods. There are many methods in this category, and we choose to discuss multiple signal classification (MUSIC) as a representative of subspace-based methods in that it is a good tradeoff between performance and computational efficiency.

Multiple input multiple output (MIMO) radar can improve the resolution of beamforming and high-resolution methods through carefully designed antenna topology and the corresponding waveforms. MIMO radar has been a research topic for a while until recently as it now finds applications to real systems such as automotive radar. In this chapter we will explain its basic working principles and common configurations and compare it with conventional phased array techniques.

The discussions of this chapter focus on techniques suitable for receive antenna array; however, when applicable many of the beamforming methods can also be used by transmit antenna array due to antenna reciprocity.

5.1 Array Manifold and Model

When antenna radiates EM waves, there are near field and far field regions around it, and the EM waves behave differently at each region. The boundary of these regions depends on the aperture (size in terms of wavelength) of the antenna. In most applications, large antennas (the size of the antenna or array is much larger than the wavelength) are being used, and the associated near/far field boundary is determined as [7]:

$$R = \frac{2D^2}{\lambda} \tag{5.1}$$

where D is the antenna length of the maximum dimension, and λ is the wavelength. Far field region is beyond the range defined by (5.1), and it is also referred to as Fraunhofer region. The near-field region can be further divided into reactive near field and radiating near field (Fresnel region). In reactive near field region, the E field and H field are not orthogonal to each other. This region is usually at the immediate vicinity of the antenna, where targets will closely couple and interact with the antenna. In fact, targets in this region can be considered part of the antenna. The Fresnel region is a transition from reactive near field to far field. The antenna beampattern can change over distance in the Fresnel region. The far field region is where most antennas operate, and we will focus our discussion on this one.

Before we introduce the concept of array manifold, let's first examine what (5.1) really means. As shown in Figure 5.1(a), the wavefront emitted from the target is a sphere. The wave arrives at the middle of the antenna array (antenna 2) with a phase different from these at the far ends of the antenna array (antenna 1 or 3) due to wave propagation distance differences. This phase delta is:

$$\Delta P = \frac{2\pi\left(\sqrt{R^2 + (D/2)^2} - R\right)}{\lambda} \tag{5.2}$$

Substitute (5.1) into (5.2), and the delta phase between the middle of the antenna and the far end is:

$$\Delta P \approx \pi / 8 \tag{5.3}$$

Therefore (5.1) or the far field boundary can be interpreted as a range beyond which the phase delta between the center of the antenna (or array) and the ends of the antenna (or array) is smaller than $\pi/8$. Under such circumstances the spherical wavefront radiated by a point target can be approximated by the plane wavefront as shown in Figure 5.1(b). In the rest of the book, we will assume that all targets are in far field region of the antenna/antenna array and the reflected waves are plane waves with wavefront traveling as a flat plane orthogonal to their propagation direction.

Let us examine a simple two-antenna array as shown in Figure 5.2(a), where the two antennas are spaced by a distance d and both receive echoes from

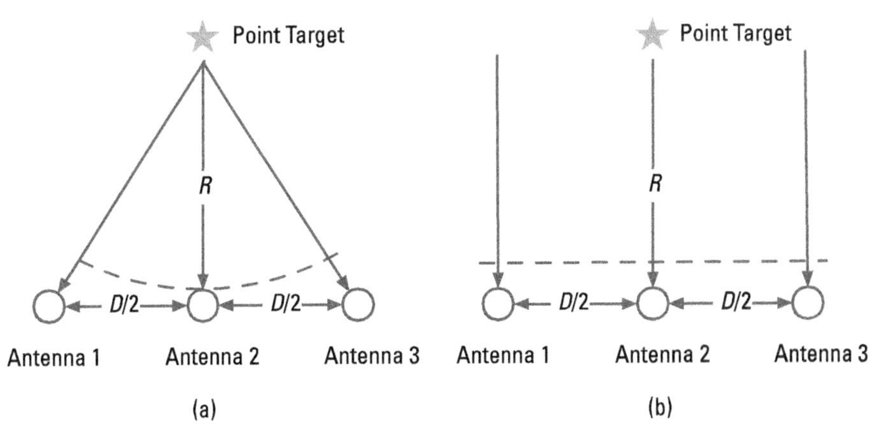

Figure 5.1 (a) Near field: wavefront propagates as a sphere; (b) far field: wavefront can be approximated by a plane, and rays from any point target are in parallel.

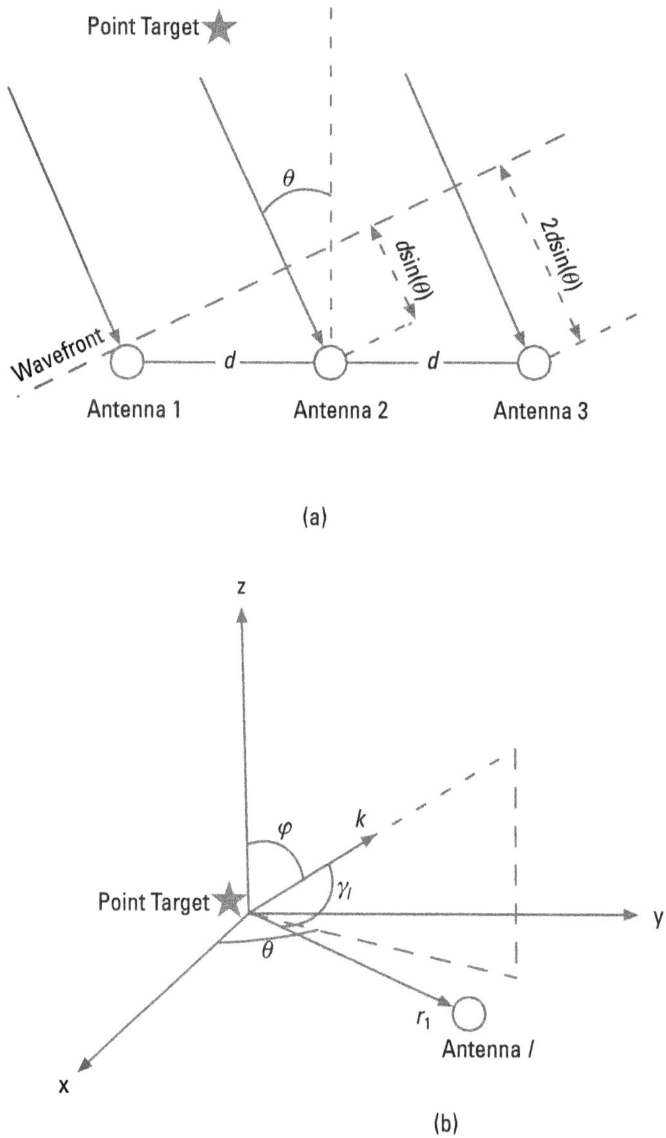

Figure 5.2 (a) Example of uniform linear array. (b) Example of antenna array with arbitrary geometry (only lth element is shown) in 3-D space. θ is defined counterclockwise relative to the x-axis in xy-plane; ϕ is defined relative to z-axis; γ_l is the angle between wavevector and the lth antenna position vector.

a point target at azimuth angle θ relative to the antenna array boresight. Based on (3.9) (the initial phases are ignored in that they are not affecting the discussion here), the signal received by antenna 1 can be expressed as:

$$s_{11}(t) = G_1(\theta) g_r(t - t_0) e^{j(2\pi f_c(t - t_0))} \tag{5.4}$$

and the signal received by antenna 2 is:

$$s_{12}(t) = G_2(\theta) g_r(t - t_0 - \tau) e^{j(2\pi f_c(t - t_0 - \tau))} \tag{5.5}$$

where f_c is the center carrier frequency and t_0 is the time delay determined by the distance between the target and antenna 1; τ is the delay that only applies to antenna 2 due to the extra distance $dsin(\theta)$ relative to antenna 1; $G_1(\theta)$ and $G_2(\theta)$ are the directivity of antenna 1 and 2, respectively. In this chapter all antennas are considered to have identical beampatterns, which is a reasonable assumption for most applications:

$$G_1(\theta) = G_2(\theta) = G(\theta) \tag{5.6}$$

Based on narrowband signal assumption (as discussed in Chapter 3), $g_r(t - t_0 - \tau) \approx g_r(t - t_0)$ due to $\tau = dsin(\theta)/c << t_0$, then (5.5) can be expressed as:

$$s_{12}(t) = G(\theta) g_r(t - t_0) e^{j(2\pi f_c(t - t_0))} e^{j(-2\pi f_c \tau)} = s_{11}(t) e^{-j2\pi dsin(\theta)/\lambda} \tag{5.7}$$

where c is speed of light, and λ is the carrier wavelength. Expression in (5.7) can be extended to an N elements uniform linear array (ULA) where the distance between any two neighboring antennas is constant; the array vector output can be expressed as:

$$\mathbf{x}(t) = [x_1(t) \quad x_2(t) \quad \cdots \quad x_N(t)]^T = G(\theta) s_{11}(t) \mathbf{a}(\theta) \tag{5.8}$$

and

$$\mathbf{a}(\theta) = \left[1 \quad e^{-\frac{j2\pi dsin(\theta)}{\lambda}} \quad e^{-\frac{j2\pi dsin(\theta)\times 2}{\lambda}} \quad \cdots \quad e^{-\frac{j2\pi dsin(\theta)(N-1)}{\lambda}} \right]^T \tag{5.9}$$

where $\mathbf{a}(\theta)$ is an $N \times 1$ vector known as steering vector and $[]^T$ represents transpose. The components of the steering vector capture the phase delta among antennas due to the slightly different propagation paths.

For an arbitrary array, these path differences can be obtained through projecting the antenna position vector to the direction of propagation. If we know the lth antenna element's position vector $r_l = [x_l, y_l, z_l]^T$ and define a wavevector **k**, which points in the direction of the wave propagation:

$$\boldsymbol{k} = k[\sin\phi\cos\theta \quad \sin\phi\sin\theta \quad \cos\phi]^T \tag{5.10}$$

the projection can be realized through the dot product of these two vectors:

$$\boldsymbol{r}_l \bullet \boldsymbol{k} = |\boldsymbol{r}_l||\boldsymbol{k}|\cos\gamma_l = k\left(x_l \sin\phi\cos\theta + y_l \sin\phi\sin\theta + z_l \cos\phi\right) \quad (5.11)$$

where γ_l is the angle between these two vectors, | | denotes modulus of a vector, and θ and ϕ are determined by the wave propagation direction as shown in Figure 5.2(b). The wavevector is not defined as a unit vector, and its magnitude $k = |\mathbf{k}| = 2\pi/\lambda$ is referred to as wavenumber that converts distance to its corresponding phase. Note the component orthogonal to the wavevector ($|\mathbf{r}_l||\mathbf{k}|\sin\gamma_l$) is ignored in that it does not contribute to the propagation path differences among antennas in plane wave scenarios. The general form of steering vector of an arbitrary array with N elements can be expressed as:

$$\mathbf{a}(\theta,\phi) = \begin{bmatrix} e^{-j\mathbf{r}_1 \bullet \mathbf{k}} & e^{-j\mathbf{r}_2 \bullet \mathbf{k}} & \cdots & e^{-j\mathbf{r}_N \bullet \mathbf{k}} \end{bmatrix}^T \quad (5.12)$$

When the array is confined in xy plane with $z_l = 0$, the steering vector simplifies into:

$$\mathbf{a}(\theta,\phi) = \begin{bmatrix} e^{-jk\left(x_1 \sin\phi\cos\theta + y_1 \sin\phi\sin\theta\right)} & \cdots & e^{-jk\left(x_N \sin\phi\cos\theta + y_N \sin\phi\sin\theta\right)} \end{bmatrix}^T \quad (5.13)$$

If the wave propagation is also limited to 2-D, which means only azimuth θ is of concern and $\phi = \pi/2$, the steering vector is:

$$\mathbf{a}(\theta) = \begin{bmatrix} e^{-jk\left(x_1 \cos\theta + y_1 \sin\theta\right)} & \cdots & e^{-jk\left(x_N \cos\theta + y_N \sin\theta\right)} \end{bmatrix}^T \quad (5.14)$$

When the antenna array is further reduced to linear array with $x_l = 0$, the steering vector is:

$$\mathbf{a}(\theta) = \begin{bmatrix} e^{-jky_1 \sin\theta} & e^{-jky_2 \sin\theta} & \cdots & e^{-jky_N \sin\theta} \end{bmatrix}^T \quad (5.15)$$

Equation (5.15) converges to (5.9) when the distance of any two neighboring elements is constant and the steering vector is normalized with respect to the first element.

After explaining the array steering vector, the general signal model for array processing can be introduced. For ease of presentation, the signal model is discussed with respect to a single real parameter referred to as the direction of arrival (DOA). However, it is straightforward to extend the established signal model and its corresponding methods to multiple parameter cases such as azimuth and elevation angles. Let us assume there are M target returns received by the antenna array from distinct DOAs $\theta_1, \theta_2, \ldots, \theta_M$, and based on superposition principle the received output vector of the array has the following form with the presence of additive noise:

$$\mathbf{x}(t) = \sum_{m=1}^{M} \mathbf{a}(\theta_m) G(\theta_m) s_m(t) + \mathbf{n}(t) \tag{5.16}$$

where $s_m(t)$ denotes the received signal waveforms from the *mth* target. n(t) is the noise vector defined as:

$$\mathbf{n}(t) = \begin{bmatrix} n_1(t) & n_2(t) & \cdots & n_N(t) \end{bmatrix}^T \tag{5.17}$$

where each component indicates the corresponding noise of the associated antenna/receiver channel. Without loss of generality, the methods in this chapter are discussed under the assumption that $\mathbf{n}(t)$ is spatially white and has the following property:

$$\mathbf{R}_\mathrm{n} = E\left\{\mathbf{n}(t)\mathbf{n}^H(t)\right\} = \sigma^2 \mathbf{I} \tag{5.18}$$

where $E\{\cdot\}$ denotes statistical expectation, $\mathbf{R}_\mathbf{n}$ is the $N \times N$ spatial covariance matrix of noise vector $\mathbf{n}(t)$, $\mathbf{I}$ is the $N \times N$ identity matrix, and $[\,]^H$ denotes conjugate transpose. Equation (5.18) indicates that all antenna elements have identical noise variance, and they are uncorrelated from each other. For the nonwhite noise situation, prewhitening techniques [2] can be applied prior to further processing. If we define a $N \times M$ steering matrix $\mathbf{A}(\theta)$ as:

$$\mathbf{A}(\theta) = \begin{bmatrix} \mathbf{a}(\theta_1) & \mathbf{a}(\theta_2) & \cdots & \mathbf{a}(\theta_M) \end{bmatrix} \tag{5.19}$$

and a vector of signal waveforms weighted by beampattern $G(\theta_m)$ as:

$$\mathbf{s}(t) = \begin{bmatrix} G(\theta_1)s_1(t) & G(\theta_2)s_2(t) & \cdots & G(\theta_M)s_M(t) \end{bmatrix}^T \tag{5.20}$$

then the received output vector (5.16) can be expressed in a more compact form:

$$\mathbf{x}(t) = \mathbf{A}(\theta)\mathbf{s}(t) + \mathbf{n}(t) \tag{5.21}$$

The covariance matrix of the received output vector is:

$$\mathbf{R} = E\left\{\mathbf{x}(t)\mathbf{x}^H(t)\right\} = \mathbf{A}(\theta)E\left\{\mathbf{s}(t)\mathbf{s}^H(t)\right\}\mathbf{A}^H(\theta) + E\left\{\mathbf{n}(t)\mathbf{n}^H(t)\right\} \tag{5.22}$$

and

$$\mathbf{P} = E\left\{\mathbf{s}(t)\mathbf{s}^H(t)\right\} \tag{5.23}$$

is the $M \times M$ covariance matrix of received signals. Combining (5.18) and (5.23) into (5.22), the covariance matrix of the output vector is:

$$\mathbf{R} = \mathbf{A}(\theta)\mathbf{P}\mathbf{A}^H(\theta) + \sigma^2\mathbf{I} \tag{5.24}$$

In the following discussions, **P** is assumed to be a nonsingular matrix. However, under certain circumstances (such as multipath where the multipath signal is a delayed and scaled version of the direct signal), the signals are coherent and result in a rank deficient **P**, and there are techniques such as spatial smoothing [8, 9] available to decorrelate the signals.

Under most circumstances covariance matrix **R** is not known, and it is a common practice to estimate it with a number of samples:

$$\hat{\mathbf{R}} = \frac{1}{L}\sum_{t=1}^{L}\mathbf{x}(t)\mathbf{x}^H(t) \tag{5.25}$$

where $L > N$ is the number of available samples for estimation, and $\hat{\mathbf{R}}$ is referred to as sample covariance matrix. When $\mathbf{x}(t)$ is white stationary Gaussian process, (5.25) turns out to be maximum likelihood estimate of **R** [2].

5.2 Conventional Beamforming

Beamforming techniques are based on the concept of steering the array to a particular direction at a time and then measuring the output power. The direction corresponding to the maximum output power is considered the DOA estimate. Steering or beamforming is achieved through linear combination of the array elements' output with a weight vector $\mathbf{w}(\theta)$:

$$y(t) = \mathbf{w}^H(\theta)\mathbf{x}(t) \tag{5.26}$$

If there are L samples available, then the average output power after steering at direction θ is:

$$P(\mathbf{w}(\theta)) = \frac{1}{L}\sum_{t=1}^{L}|y(t)|^2 = \frac{1}{L}\sum_{t=1}^{L}y(t)y^*(t) = \mathbf{w}^H(\theta)\hat{\mathbf{R}}\mathbf{w}(\theta) \tag{5.27}$$

The conventional beamformer (also known as Bartlett beamformer) maximizes the output power for an input signal at a given direction (Chapter 2 of [1]). Let us consider a return signal arriving from direction $\tilde{\theta}$, then the array output is:

$$\mathbf{x}(t)=\mathbf{a}(\tilde{\theta})s(t)+\mathbf{n}(t) \tag{5.28}$$

The optimization problem can be formulized as:

$$\max_{\mathbf{w}(\theta)} E\left\{\mathbf{w}^H(\theta)\mathbf{x}(t)\mathbf{x}^H(t)\mathbf{w}(\theta)\right\}=\max_{\mathbf{w}(\theta)} \mathbf{w}^H(\theta)E\left\{\mathbf{x}(t)\mathbf{x}^H(t)\right\}\mathbf{w}(\theta) \tag{5.29}$$

Substituting (5.28) into (5.29) and considering the zero mean noise vector is uncorrelated to the signal, the following is true:

$$\begin{aligned}&\max_{\mathbf{w}(\theta)} E\left\{\mathbf{w}^H(\theta)\mathbf{x}(t)\mathbf{x}^H(t)\mathbf{w}(\theta)\right\}\\&=\max_{\mathbf{w}(\theta)}\left\{E\left\{|s(t)|^2\right\}\left|\mathbf{w}^H(\theta)\mathbf{a}(\tilde{\theta})\right|^2+\sigma^2|\mathbf{w}(\theta)|^2\right\}\end{aligned} \tag{5.30}$$

The norm of $\mathbf{w}(\theta)$ has to be constrained to avoid trivial solution, and we can set $|\mathbf{w}(\theta)|^2 = 1$ for convenience. This optimization problem then simplifies into:

$$\max_{\mathbf{w}(\theta)} E\left\{\mathbf{w}^H(\theta)\mathbf{x}(t)\mathbf{x}^H(t)\mathbf{w}(\theta)\right\}=\max_{\mathbf{w}(\theta)}\left\{E\left\{|s(t)|^2\right\}\left|\mathbf{w}^H(\theta)\mathbf{a}(\tilde{\theta})\right|^2+\sigma^2\right\} \tag{5.31}$$

The Cauchy-Schwarz inequality states that for any two vectors the following is always true:

$$\left|\mathbf{w}^H(\theta)\mathbf{a}(\tilde{\theta})\right|^2\le|\mathbf{w}(\theta)|^2\left|\mathbf{a}(\tilde{\theta})\right|^2 \tag{5.32}$$

and the two sides are equal if and only if $\mathbf{w}(\theta)$ is a scaled version of (linearly dependent on) $\mathbf{a}(\tilde{\theta})$. Considering $|\mathbf{w}(\theta)|^2 = 1$, the solution of (5.31) is:

$$\mathbf{w}(\theta)=\frac{\mathbf{a}(\tilde{\theta})}{\sqrt{\mathbf{a}^H(\tilde{\theta})\mathbf{a}(\tilde{\theta})}} \tag{5.33}$$

The weight vector (5.33) effectively aligns the phases of each antenna element prior to linearly combining them to achieve maximum power. The resulting signal-to-noise ratio improvement is equal to the number of antenna elements N, which is referred to as array processing gain or beamforming gain.

Considering (5.33) and (5.27), the spatial spectrum of Bartlett beamformer is:

$$P(\mathbf{w}(\theta)) = \frac{\mathbf{a}^H(\theta)\hat{\mathbf{R}}\mathbf{a}(\theta)}{\mathbf{a}^H(\theta)\mathbf{a}(\theta)} \tag{5.34}$$

with θ being any possible value of the DOA. The true DOA of targets can then be identified from the peaks in the resulting spatial spectrum of the beamformer.

5.2.1 Uniform Array and FFT Based Beamforming

For the most commonly used ULA, the weight vector has the following form according to (5.9) and (5.33):

$$\mathbf{w}_{ULA}(\theta) = \begin{bmatrix} 1 & e^{-\frac{j2\pi d sin(\theta)}{\lambda}} & e^{-\frac{j2\pi d sin(\theta)\times 2}{\lambda}} & \cdots & e^{-\frac{j2\pi d sin(\theta)(N-1)}{\lambda}} \end{bmatrix}^T / \sqrt{N} \tag{5.35}$$

The corresponding Bartlett beamformer output is:

$$y(t) = \mathbf{w}^H(\theta)\mathbf{x}(t) = \sum_{i=1}^{N} e^{-\frac{j2\pi d sin(\theta)}{\lambda}(i-1)} x_i(t)/\sqrt{N} = \sum_{i=1}^{N} e^{-j2\pi\Omega(\theta)(i-1)} x_i(t)/\sqrt{N} \tag{5.36}$$

where

$$\Omega(\theta) = d sin(\theta)/\lambda \tag{5.37}$$

can be treated as spatial frequency normalized against sampling frequency of 1. Equation (5.36) is simply a normalized discrete time Fourier transform applied to the sensor array data. All the Fourier spectral analysis theories and limitations in time domain can be equally applied here. Nyquist theory states that the maximum frequency in the signal needs to be less than half of the sampling frequency in order to avoid aliasing (manifest as grating lobes in the spectrum). Considering (5.37) and sampling frequency of 1, the following condition needs to be satisfied to avoid grating lobes:

$$d\left|sin(\theta)\right|/\lambda \le 1/2 \Rightarrow d \le \lambda/\left(2\left|sin(\theta)\right|\right) \tag{5.38}$$

If the array coverage is up to 90 degrees, then the maximum distance between two neighboring elements is $\lambda/2$ without grating lobes. In practice, the individual antenna beampattern can limit the field of view to be less than +/–90 degrees, and the distance between antenna elements can be increased slightly based on (5.38).

In many radar systems such as automotive radar, FFT is often adopted as the implementation algorithm for Bartlett beamformer due to its computational efficiency. This is the case especially when there are multiple beams (>16) formed. If there are K beams formed on the N elements array, the kth FFT bin can be expressed as:

$$X_k = \sum_{i=1}^{N} e^{-j2\pi(i-1)k/K} x_i(t) \qquad 0 \le k \le K-1 \tag{5.39}$$

where $\frac{k}{K} = d\sin(\theta)/\lambda$ and the mapping between angles and FFT bins is:

$$\begin{cases} \theta = \sin^{-1}\left(\dfrac{k\lambda}{Kd}\right) & for\, 0 \le k < K/2 \\ \theta = \sin^{-1}\left(\dfrac{(k-K)\lambda}{Kd}\right) & for\, \dfrac{K}{2} \le k \le K-1 \end{cases} \tag{5.40}$$

The FFT bins to DOA angle mapping is plotted in Figure 5.3 for an example 32 element ULA with 64 beams formed. The element distance is selected to be $d = \lambda/2$. It is worth noting that when applying FFT as the Bartlett

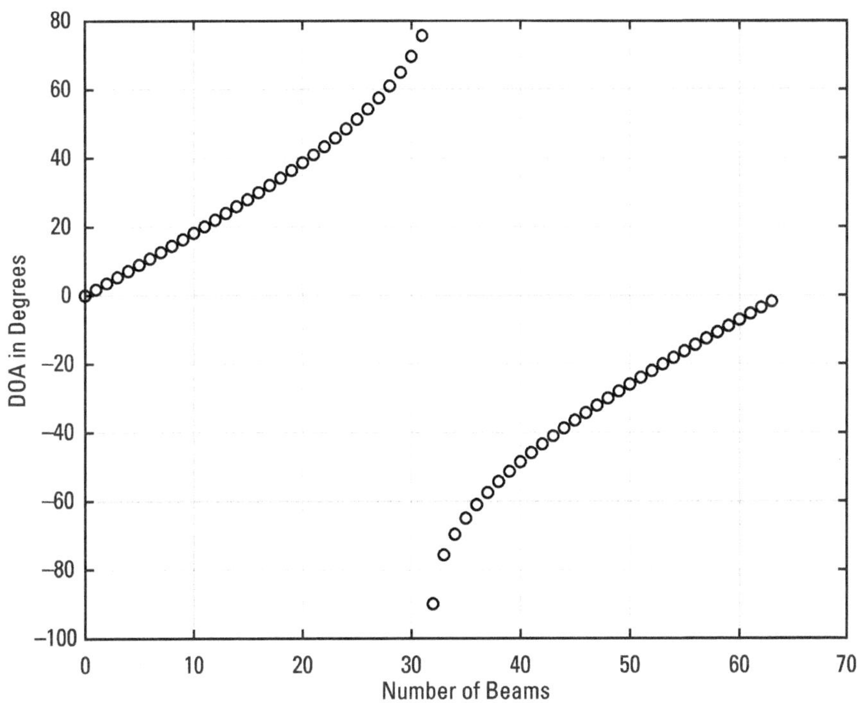

Figure 5.3 The DOAs (in degrees) versus the corresponding FFT beams (bins).

beamformer, the formed beams are not uniform. The beamformer has finer step size near zero degrees (boresight) and coarser ones toward both ends.

The Bartlett beamformer is widely used because it is not only efficient but also robust. There is no prior knowledge required on the signal model and the number of signals. It works with nonwhite noise and singular **P** matrix.

For FMCW radars, there are possibly three FFTs that correspond to fast time processing, slow time processing, and beamforming, respectively. After processing these, a data cube is formed, as shown in Figure 5.4.

5.2.2 Array Resolution, Accuracy, and Sidelobes Control

Since the conventional beamformer is effectively Fourier analysis extension to the spatial domain, it possesses the same resolution and accuracy limitations. This means that the spatial frequency $\Omega(\theta)$ has a resolution of $1/N$ after beamforming. For the convenience of discussion, let us assume that the beam is scanned to angle θ_0, and it is corresponding to the center of a spatial frequency resolution bin, and θ_+ is at the upper end of the same resolution bin. Then the difference of their corresponding spatial frequencies is half of the resolution bin:

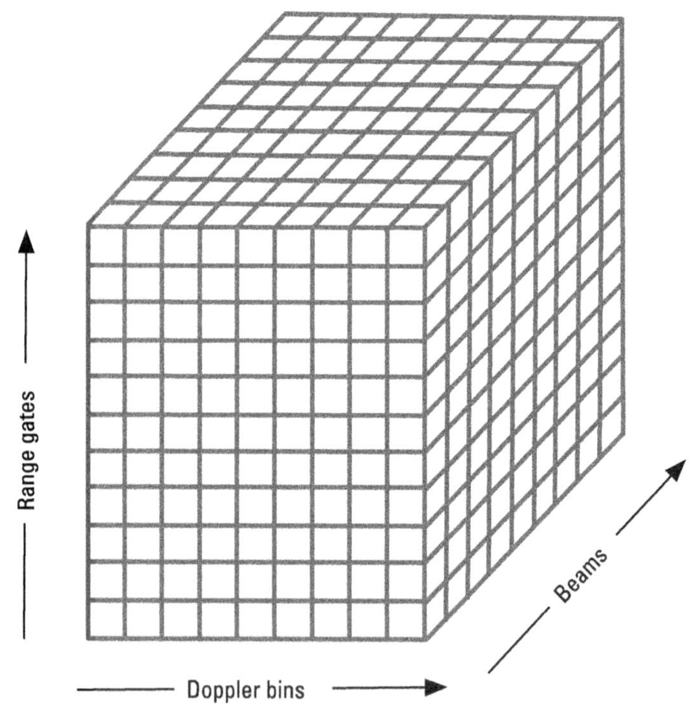

Figure 5.4 Data cube after fast time, slow time, and beamforming have been applied to the received data.

$$\frac{d\sin(\theta_+)}{\lambda}-\frac{d\sin(\theta_0)}{\lambda}=\frac{1}{2}\frac{1}{N} \tag{5.41}$$

which means:

$$\sin(\theta_+)-\sin(\theta_0)=\frac{\lambda}{2Nd} \tag{5.42}$$

The left side of (5.42) can be expressed as:

$$\sin(\theta_+)-\sin(\theta_0)=\sin(\theta_+-\theta_0)\cos(\theta_0)+\cos(\theta_+-\theta_0)\sin(\theta_0)-\sin(\theta_0) \tag{5.43}$$

When $\theta_+ - \theta_0$ is small, (5.43) can be approximated by:

$$\sin(\theta_+)-\sin(\theta_0)\approx(\theta_+-\theta_0)\cos(\theta_0) \tag{5.44}$$

Combining (5.42) and (5.44), the angle resolution or beamwidth is:

$$\theta_B=2(\theta_+-\theta_0)=\frac{\lambda}{Nd\cos(\theta_0)} \tag{5.45}$$

Note the 3-dB beamwidth ($\theta_{3\text{dB}} = 0.886\theta_B$) is slightly different from (5.45) as explained in [10]. The angle measurement accuracy of a uniform array has been shown to be [11]:

$$\delta\theta=\frac{0.628\theta_{3dB}}{(2E/N_0)^{1/2}} \tag{5.46}$$

where E/N_0 is the signal-to-noise ratio with the array processing gain. It is obvious that the angle resolution and accuracy are depending on the pointing angle.

Similar to Sections 4.1.5 and 4.2.7, there are sidelobes at –13.2 dB with respect to the main lobe in the corresponding Bartlett beamformer outputs. The returns from weak targets may be masked by the sidelobes of strong targets or there may be ghost targets due to sidelobe breakthrough. Applying window functions prior to beamforming is a common practice, and the design of window function is also analogous to that applied to spectral leakage in Fourier analysis. The readers can refer to Section 4.1.5 on the tradeoff and how to select the window functions.

5.2.3 Digital Beamforming Versus Analog Beamforming

Analog beamforming applies weight vectors in an analog way (analog phase shifters or time delay modules) to combine the signals from the antenna array. The analog beamformer requires only one RF channel and one ADC, and it can only form one beam at a time. Digital beamforming, on the other hand, requires each antenna element to have its own RF channel and ADC to digitize the signal. Simultaneous multiple beams are formed in the digital processor. When the number of antenna elements is large, the associated hardware cost of a digital beamformer can be prohibitive. The structure of a digital beamformer is more flexible. Advanced signal processing techniques such as adaptive beamforming, MIMO, and high-resolution methods can be applied due to the availability of data from all antenna elements.

5.3 High-Resolution Methods

In order to alleviate the limitations of conventional beamformer, there are many methods proposed in the literature. Adaptive beamforming selects the weight vector based on the data covariance matrix, which can adaptively set the nulls in the resultant beampattern to better reduce strong interference while also improving the resolution. Maximum likelihood methods can significantly improve the resolution at the cost of computational resources due to multidimensional search for the solution. Subspace-based methods are good tradeoff between computation intensity and resolution. In this section MUSIC is explained as a representative of subspace-based methods, which has been applied in automotive radars. Readers of interest can refer to [1, 2, 4, 5] for further treatment of variations of MUSIC and other high-resolution methods. Capon's beamformer is also briefly discussed as a representative of adaptive beamforming due to its wide application in practice.

MUSIC

The covariance matrix (5.24) is Hermitian and positive definite according to its definition in (5.22), which guarantees the following decomposition [12]:

$$\mathbf{R} = \mathbf{A}(\theta)\mathbf{P}\mathbf{A}^{H}(\theta) + \sigma^2\mathbf{I} = \mathbf{U}\boldsymbol{\Lambda}\mathbf{U}^{H} \tag{5.47}$$

where $\mathbf{U}$ is a unitary matrix consisting of eigenvectors of $\mathbf{R}$. $\boldsymbol{\Lambda}$ is a diagonal matrix with real eigenvalues of $\mathbf{R}$ in a descending order:

$$\Lambda = \begin{bmatrix} \lambda_1 & 0 & \cdots & 0 \\ 0 & \lambda_2 & \cdots & 0 \\ \vdots & \vdots & \ddots & \vdots \\ 0 & 0 & \cdots & \lambda_N \end{bmatrix} \tag{5.48}$$

and

$$\lambda_1 \geq \lambda_2 \geq \cdots \geq \lambda_N > 0 \tag{5.49}$$

MUSIC requires the number of targets M in the received signals to be smaller than the number of array element N. According to (5.24), any vector $\mathbf{u}_n$ which is orthogonal to the columns of $\mathbf{A}(\theta)$ is an eigenvector of $\mathbf{R}$ with the corresponding eigenvalue σ^2. Because $\mathbf{A}(\theta)$ has a rank of M, there are $N - M$ such linearly independent eigenvectors whose eigenvalues are equal to σ^2:

$$\lambda_{M+1} = \lambda_{M+2} = \cdots = \lambda_N = \sigma^2 \tag{5.50}$$

These $N - M$ eigenvectors/eigenvalues are known as noise eigenvectors/eigenvalues and the space spanned by them is referred to as noise subspace. If vector $\mathbf{u}_s$ is an eigenvector of matrix $\mathbf{A}(\theta)\mathbf{P}\mathbf{A}^H(\theta)$ with the corresponding eigenvalue σ_s^2, then the following is true:

$$\mathbf{R}\mathbf{u}_s = \mathbf{A}(\theta)\mathbf{P}\mathbf{A}^H(\theta)\mathbf{u}_s + \sigma^2\mathbf{I}\mathbf{u}_s = (\sigma_s^2 + \sigma^2)\mathbf{u}_s \tag{5.51}$$

which means $\mathbf{u}_s$ is also an eigenvector of $\mathbf{R}$ with corresponding eigenvalue $(\sigma_s^2 + \sigma^2)$. Because $\mathbf{A}(\theta)\mathbf{P}\mathbf{A}^H(\theta)$ has a rank of M, there are M such linearly independent eigenvectors and their corresponding eigenvalues are larger than σ^2. These eigenvectors/eigenvalues are known as signal eigenvectors/eigenvalues and the space spanned by them is referred to as signal subspace.

Based on the previous discussion, (5.47) can be rewritten as:

$$\mathbf{R} = \begin{bmatrix} \mathbf{U}_s & \mathbf{U}_n \end{bmatrix} \begin{bmatrix} \Lambda_s & 0 \\ 0 & \Lambda_n \end{bmatrix} \begin{bmatrix} \mathbf{U}_s^H \\ \mathbf{U}_n^H \end{bmatrix} = \mathbf{U}_s\Lambda_s\mathbf{U}_s^H + \mathbf{U}_n\Lambda_n\mathbf{U}_n^H \tag{5.52}$$

where

$$\Lambda_n = \sigma^2\mathbf{I} \tag{5.53}$$

The $N \times M$ matrix $\mathbf{U}_s$ is composed of the M signal eigenvectors, and the $N \times (N - M)$ matrix $\mathbf{U}_n$ consists of these noise eigenvectors. The $M \times M$ matrix Λ_s is a diagonal matrix with diagonal elements being the corresponding signal eigenvalues. Since the noise eigenvectors are orthogonal to the matrix $\mathbf{A}(\theta)$, the following is true:

$$\mathbf{U}_n^H \mathbf{a}(\theta) = 0, \theta \in \{\theta_1, \theta_2, \cdots \theta_M\} \tag{5.54}$$

When $N > M$ and $\mathbf{P}$ is full rank, the only possible solutions to (5.54) are θ_1, θ_2, ..., θ_M. Therefore, the MUSIC spatial spectrum can be defined as the following to locate DOAs:

$$P_{MUSIC}(\theta) = \frac{\mathbf{a}^H(\theta)\mathbf{a}(\theta)}{\mathbf{a}^H(\theta)\mathbf{U}_n\mathbf{U}_n^H\mathbf{a}(\theta)} \tag{5.55}$$

In practice, we can have the following steps to obtain the DOAs:

1. Obtain the sample covariance matrix $\hat{\mathbf{R}}$ through (5.25).
2. Conduct the eigen-decomposition on $\hat{\mathbf{R}}$ through (5.47):

$$\hat{\mathbf{R}} = \hat{\mathbf{U}}\hat{\Lambda}\hat{\mathbf{U}}^H$$

3. Take the last $N - M$ columns of $\hat{\mathbf{U}}$ as the noise subspace $\hat{\mathbf{U}}_n$. (The number of underlying signals M is considered known; however, there are many effective methods [13, 14] on how to estimate it when unknown.)
4. Obtain the spatial spectrum as in (5.55) and locate the peaks in the spectrum as estimate of DOAs.

The performance of MUSIC can be significantly better than that of beamforming. In fact, the theoretical MUSIC's performance can be arbitrarily good as long as there are long enough data or a high enough signal-to-noise ratio with sufficiently accurate signal model. However, in practice MUSIC performance is bounded by realistic data limitation or inaccurate underlying signal model such as steering vector variations, rank deficient P matrix, unknown target number, or colored noise.

The simulation results are shown in Figure 5.5, where there is a 16-element ULA spaced at half wavelength and there are two targets in the field of view with DOAs of –20 and –25 degrees, respectively. The signal-to-noise ratio is 10 dB for both targets. The solid trace represents the spatial spectrum of MUSIC, where the two peaks correspond to the correct DOAs. The Bartlett beamformer cannot resolve these targets in that there is only one peak around –22 degrees in the spectrum as indicated by the dashed trace.

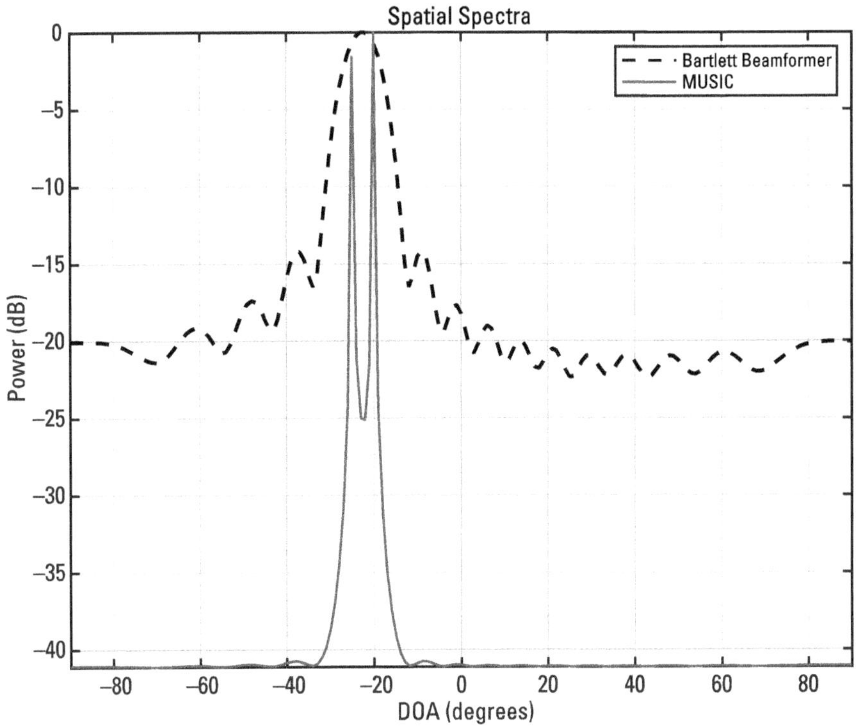

Figure 5.5 Spatial spectrum of MUSC (solid trace) and Bartlett beamformer (dashed trace).

Capon Beamformer

Capon beamformer is also known as minimum variance distortionless response (MVDR), which can be found in many real-world applications due to its capability to suppress strong interference. MVDR is the solution of the following optimization problem:

$$\begin{aligned}&\underset{\mathbf{w}(\theta)}{\text{Min}}\, P\left(\mathbf{w}(\theta)\right)\\&\text{subject to } \mathbf{w}^H(\theta)\mathbf{a}(\theta)=1\end{aligned} \tag{5.56}$$

where $p(\mathbf{w}(\theta))$ is defined in (5.27). Solution of (5.56) can be considered a sharp spatial band pass filter that maintains a fixed gain in the desired (look) direction while minimizing all noise and signals from other directions. Equation (5.56) can be solved through Lagrange multipliers with the solution as [1]:

$$\mathbf{w}(\theta)=\frac{\hat{\mathbf{R}}^{-1}\mathbf{a}(\theta)}{\mathbf{a}^H(\theta)\hat{\mathbf{R}}^{-1}\mathbf{a}(\theta)} \tag{5.57}$$

Considering (5.57) and (5.27), the spatial spectrum of MVDR is:

$$P_{MVDR}(\theta) = \frac{1}{\mathbf{a}^H(\theta)\hat{\mathbf{R}}^{-1}\mathbf{a}(\theta)} \tag{5.58}$$

The minimization process in (5.56) will automatically put nulls at the directions of strong interference. Similarly, when there are closely spaced targets, the leakage from the other sources are reduced, and therefore MVDR has better resolution than that of the Bartlett beamformer.

5.4 MIMO

MIMO radar has recently attracted significant research activity for its resolution advantages, which led to a multitude of proposed algorithms, architectures, and associated complex theories. MIMO radar transmits mutually orthogonal waveforms from multiple transmit antennas, and the returns from these waveforms are separated by each of the receive antennas. After rearranging the data, a virtual extended array can be formed, which is much bigger than either the transmit array or the receive array. There is a large number of references addressing specific aspects of the MIMO radar from waveform design, overall architecture to performance analysis with extensive mathematics derivations. Due to space limitation, we will focus our discussions to the fundamentals of MIMO radar and the common adopted waveforms that can be found in real-life applications. We will only discuss coherent MIMO radar with colocated transmit and receive antennas, which has been recently adopted in civil radars. For more extended and in-depth presentations, the reader is referred to references such as [3, 15].

5.4.1 Virtual Array

The fundamentals of MIMO can be explained through the concept of virtual array. We will elaborate on this concept with a simple three element receive ULA and two distributed transmit antennas, as shown in Figure 5.6. The two transmit antennas are on both ends of the receive array with a distance of $3d$. The two transmits antennas are assumed to have identical beampattern and maintain phase coherency. According to (5.4), the waveform sent from *Tx 1* is reflected by the target, and the echoes are received by *Rx 1*:

$$s_{11}(t) = G(\theta)g_r(t - t_0)e^{j(2\pi f_c(t-t_0))} \tag{5.59}$$

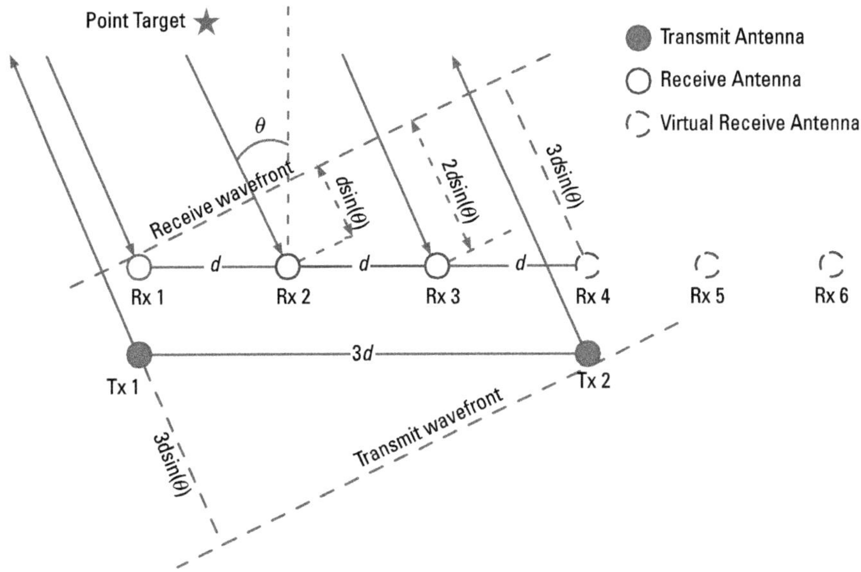

Figure 5.6 Illustration of virtual array. Solid circles: transmit antennas. Circles: receive antennas. Dashed circles: virtual receive antennas.

According to (5.7), the same target returns received by *Rx 2* and *Rx 3* are, respectively:

$$s_{12}(t) = s_{11}(t)e^{-j2\pi\frac{dsin(\theta)}{\lambda}} \tag{5.60}$$

and

$$s_{13}(t) = s_{11}(t)e^{-j2\pi\frac{2dsin(\theta)}{\lambda}} \tag{5.61}$$

The waveform sent from *Tx 2* is reflected by the same target and received by *Rx 1*:

$$s_{21}(t) = G(\theta)g_r(t - t_0 - \tau)e^{j(2\pi f_c(t-t_0-\tau))} \tag{5.62}$$

where t_0 is the time delay determined by the distance from *Tx* 1 to the target and back to *Rx* 1; τ is the time delay that applies to *Tx* 2 returns due to the extra distance $3d\sin(\theta)$.

This extra distance is the difference between the signal propagation path from *Tx 1* to the target and that from *Tx 2* to the target. Considering the narrowband nature, (5.62) can be expressed as:

$$s_{21}(t) = G(\theta)g_r(t - t_0)e^{j(2\pi f_c(t-t_0))}e^{-j2\pi f_c 3dsin(\theta)/c} = s_{11}(t)e^{-j2\pi\frac{3dsin(\theta)}{\lambda}} \tag{5.63}$$

If there is a fourth virtual receive antenna (*Rx 4* as shown in Figure 5.6), then the *Tx 1* related echoes received by *Rx 4* will have the same expression as (5.63). We can conclude that when transmitting from *Tx 2* and receiving at *Rx 1* is equivalent to transmitting from *Tx 1* and receiving at *Rx 4*. By analogy, the same applies to *Rx 2* and *Rx 3*. From the received signals point of view, *Tx 2/Rx 2* and *Tx 2/Rx 3* are equivalent to *Tx 1/Rx 5* and *Tx 1/Rx 6*, respectively. Thereafter the two transmit antenna and three receive antenna array structure can be treated as a single transmit antenna and six receive antennas structure. The effective array aperture is significantly increased. Due to reciprocity, the roles of transmit and receive antennas can be exchanged, and the final virtual array stays the same.

These discussions can be generalized to a Q transmit antenna array and N receive antenna array, and the equivalent virtual array is the convolution of these two arrays resulting in QN element virtual array. The virtual array formation can be further extended to nonuniform planar array as an example shown in Figure 5.7. The sparse array can be either transmit array or receive array.

After rearranging received data and forming an effective virtual array, the previously discussed array processing techniques such as beamforming or MUSIC can be applied. It is worth pointing out that the near/far field boundary formula (5.1) should take into account the virtual array size instead of the physical array size when MIMO techniques are deployed. This is also the case for angle resolution and accuracy formulas (5.45) and (5.46).

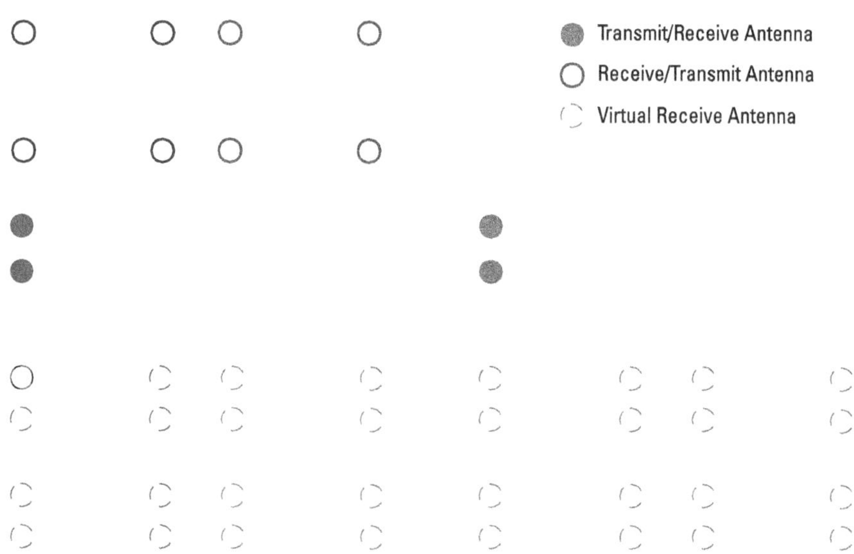

Figure 5.7 Virtual array of two nonuniform planar arrays.

5.4.2 Basic MIMO Waveforms

In Section 5.4.1 we assume ideal orthogonal waveforms and each receive antenna can separate returns from different transmit antennas. In practice, how to choose proper and practical waveforms is critical to the success of MIMO radar. There are four categories of well-known waveforms: time division multiple access (TDMA), frequency division multiple access (FDMA), Doppler division multiple access (DDMA), and code division multiple access (CDMA). In this section we will briefly discuss TDMA, DDMA, and CDMA, as these are commonly adopted in civil radar systems.

TDMA

In TDMA, each transmit antenna sends out its waveforms alternatively (i.e., there is no overlap in any two transmissions). TDMA is straightforward to apply, has perfect orthogonality, and works with any conventional waveforms. However, it has two shortcomings. The first one is that at any time there is only one transmit antenna working, which results in significant loss of transmitter power and coverage. The second one is that from one transmission to the next transmission, phase changes are induced from relative motion between radar and targets.

For LFMCW radar, there is a solution [16, 17] to address the transmitter power deficiency. All transmit antennas are active nearly the same time but with the start time slightly staggering one after another. An example is shown in Figure 5.8 for four transmit antennas. The returns from different transmit antennas will center around different beat frequencies and can be readily separated by each receive antenna. The negative impact of this solution is the reduction of the maximum unambiguous range by a factor equal to the number of transmit antennas.

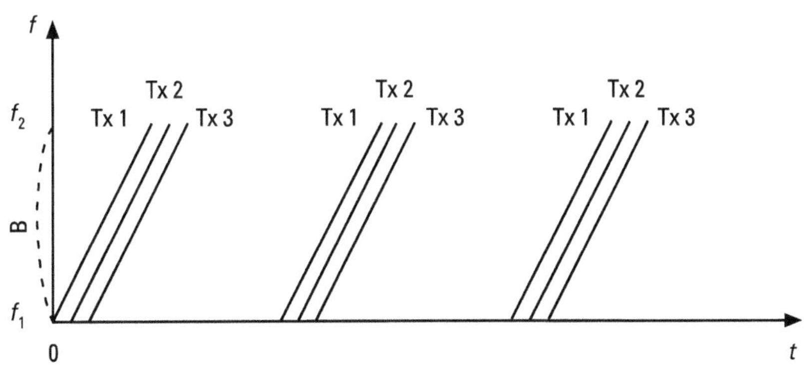

Figure 5.8 Example of staggered LFM waveforms for TDMA MIMO with three transmit antennas.

To compensate motion-related phase changes, [18] proposes an overlapping antenna in the virtual array of any two consecutive transmitting antennas. If these two transmissions happened simultaneously, the phases of the overlapping antenna should be the same for both transmissions. Therefore, the phase difference of these two transmissions at the overlapping antenna can be used to correct the phase changes. However, this method requires strong SNR and suffers loss of $Q-1$ elements in the final virtual array due to overlapping antennas. To overcome these disadvantages, [19] presents a modified DFT method that can compensate the phase changes for each Doppler component of the received signal without requiring overlap antennas.

DDMA

The idea of DDMA MIMO is similar to that of FDMA MIMO but has much less demand on the overall bandwidth. Under most circumstances, the frequency bandwidth is precious and constrained by regulations, and DDMA is more attractive than FDMA in practice. For fast time (within a single waveform) DDMA sends slightly different center frequency from each transmit antenna and the received signals can be separated in Doppler domain. The center frequency offset of any two transmit antennas needs to be larger than twice of the maximum Doppler shift, which is generally much smaller than the offset (the waveform bandwidth) in FDMA. In slow time (across chirp/pulse) implementation of DDMA, the center frequency offset can be achieved through modulating the initial phase of each chirp/pulse. Many candidate phase coding schemes are available, such as the binary Hadamard code as explained in [17]. In this section, we will use Frank poly phase code [20] as an example to explain how to achieve slow time orthogonality in Doppler domain.

Let us assume there are Q transmit antennas and X chirps/pulses in the CIT window. The corresponding Frank code has $X \times X$ element and the i subcode's j^{th} element is:

$$\phi_{i,j} = \frac{2\pi}{X}(i-1)(j-1), \quad 1 \le i \le X, 1 \le j \le X \tag{5.64}$$

The waveforms at the qth transmit antenna is modulated by the ith subcode with $i = \frac{(q-1)X}{Q} + 1$, which means the initial phases of the waveforms are determined by the elements of the corresponding subcode. Let us assume there is an arbitrary target at range gate n. Then for the xth waveform sent from the first transmit antenna, the received signals have the following expression according to (4.13):

$$s_1(n,x) = Ae^{j\theta_0} e^{j\frac{2\pi \bullet 2v \bullet t_{wri} \bullet (x-1)}{\lambda}} \qquad 1 \le x \le X \tag{5.65}$$

where t_{wri} is the time between two neighboring waveforms. Equation (5.65) has corresponding Doppler spectrum of $S_1(n, f)$ as defined in (4.14). The returns associated with the qth transmit antenna is:

$$s_q(n,x) = e^{j\frac{2\pi}{X}\frac{(q-1)X}{Q}(x-1)} Ae^{j\theta_0} e^{j\frac{2\pi \bullet 2v \bullet t_{wri} \bullet (x-1)}{\lambda}} \qquad 1 \le x \le X \tag{5.66}$$

and its corresponding Doppler spectrum is:

$$S_q(n,f) = \delta\left(f - \frac{f_{wrf}(q-1)}{Q}\right) * S_1(n,f) \tag{5.67}$$

where * indicates convolution and f_{wrf} is the waveform repetition frequency. Equation (5.67) states that the Doppler spectrum associated with the qth transmit antenna is shifted in the Doppler domain by $\frac{f_{wrf}(q-1)}{Q}$. Therefore, the signal returns from different transmit antennas can be separated if the maximum Doppler frequency is less than $f_{wrf}/(2Q)$. The tradeoff of DDMA is obviously the reduction of maximum unambiguous Doppler frequency from $f_{wrf}/2$ to $f_{wrf}/(2Q)$.

CDMA

In CDMA, the waveforms transmitted by different antennas are modulated by a different set of pseudorandom phase codes, and these phases codes are orthogonal to each other. CDMA can be realized either in fast time or in slow time. The fast time implementation requires each waveform to be modulated by the codes at carrier frequency, which sets a high demand on hardware. It has been discussed in Section 3.4. In slow time CDMA implementation, the phase codes are used to modulate the initial phases of different waveforms in the same way as discussed earlier in slow time DDMA. The only difference is that the pseudorandom phase codes are used instead of Frank polyphase code. The slow time CDMA can be implemented with the same transceiver architecture as traditional FMCW radar; however, the achievable dynamic range is limited by the codes sequence length X (or equivalently the number of waveforms within a CIT window) and the number of transmit antennas Q:

$$dR = 10log_{10}\left(X/(Q-1)\right) \tag{5.68}$$

5.4.3 Summary

MIMO techniques can help improve the angle resolution and accuracy through forming a virtual array. The waveforms are key to enable MIMO, and there is no universal solution. The modulation schemes of DDMA, TDMA, FDMA, and CDMA are summarized in Figure 5.9, where the ways to separate the various transmitted waveforms are visually explained. Each method has its own pros and cons, and the waveform selection needs to be based on real-life radar system requirements.

It is also worth pointing out that MIMO is more flexible and has better angle resolution than conventional phased array techniques; however, MIMO suffers loss in SNR relative to phased arrays. This is because the transmit antennas in MIMO forms noncoherent beams while phased array has coherent formation of beams. Therefore, engineers may choose different technologies based on particular use cases.

The array signal processing is an important field and accumulated many years' knowledge and advanced algorithms. However, in some gesture recognition applications, the machine learning, especially deep neural net, provides better classification results with raw channel data compared to that after array signal processing. These interesting findings will be discussed in later chapters.

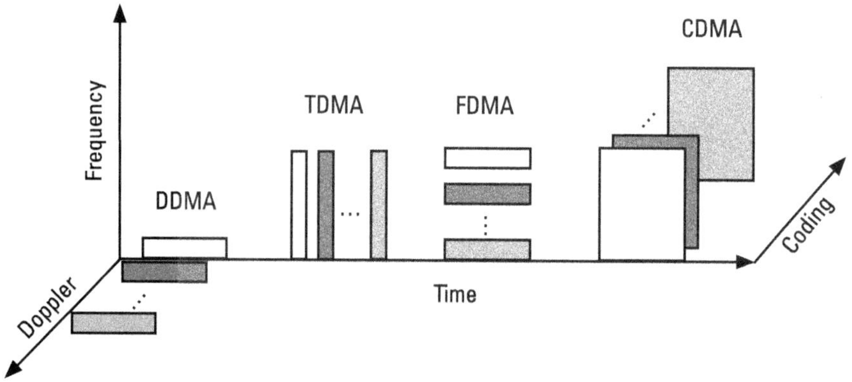

Figure 5.9 Illustration of DDMA, TDMA, FDMA, and CDMA distribution in time, frequency, Doppler, and coding space.

References

[1] Van Trees, H. L., *Optimum Array Processing—Part IV of Detection, Estimation, and Modulation Theory*, New York: John Wiley, 2002.

[2] Krim, H., and M. Viberg, "Two Decades of Array Signal Processing Research,"' *IEEE Transactions on Signal Processing Magazine*, July 1996, pp. 67–94.

[3] Li, L., and P. Stoica, *MIMO Radar Signal Processing*, Hoboken, NJ: John Wiley & Sons, 2009.

[4] Zourbir, A. M., et al., *Academic Press Library in Signal Processing, Vol 3, Array and Statistical Signal Processing*, Chapter 20, Chennai: Academic Press, 2014.

[5] Benesty, J., et al., *Fundamentals of Signal Enhancement and Array Signal Processing*, Hoboken, NJ: John Wiley & Sons, 2018.

[6] Chandran, S., *Adaptive Antenna Arrays Trends and Applications*, Berlin: Springer Verlag, 2004.

[7] Balanis, C. A., *Antenna Theory Analysis and Design*, Chapter 2, Hoboken, NJ: John Wiley & Sons, 2005.

[8] Shan, T. J., et al., "On Spatial Smoothing for Directions of Arrival Estimation of Coherent Signals," *IEEE Trans. on Acoustics, Speech and Signal Processing*, Vol. 33, No. 4, August 1985, pp. 806–811.

[9] Friedlander, B., and A. Weiss, "Direction Finding Using Spatial Smoothing with Interpolated Arrays," *IEEE Trans. on AES*, Vol 28, No. 2, April 1992, pp. 574–587.

[10] Skolnik, M. I., *Introduction to Radar Systems*, Third Edition, Chapter 9.6, New York: McGraw-Hill, 2001.

[11] Skolnik, M. I., *Introduction to Radar Systems*, Third Edition, Chapter 6.4, New York: McGraw-Hill, 2001.

[12] Lay, D. C., et al., *Linear Algebra and Its Applications*, Chapter 7, Harlow: Pearson Education Limited, 2016.

[13] Wax, M., et al., "Spatio-Temporal Spectral Analysis by Eigenstructure Methods," *IEEE Trans. on ASSP*, ASSP-32, Aug. 1984.

[14] Cheng, W., et al., "A Comparative Study of Information-Based Source Number Estimation Methods and Experimental Validations on Mechanical Systems," *Sensors*, 14, 2014, pp. 7625–7646.

[15] Bergin, J., and J. R. Guerci, *MIMO Radar Theory and Application*, Boston, MA: Artech House, 2018.

[16] Frazer, G. J., et al., "Recent Results in MIMO Over-the-Horizon Radar," *2008 IEEE Radar Conference*, Italy, May 2008, pp. 789–794.

[17] Sun, H., et al., "Analysis and Comparison of MIMO Radar Waveforms," *International Radar Conference*, 2014, pp. 1–6.

[18] Schmid, C., et al., "Motion Compensation and Efficient Array Design for TDMA FMCW MIMO Radar Systems," *Sixth European Conference on Antennas and Propagation* (EUCAP), Mar. 2012), pp. 1746–1750.

[19] Bechter, J., et al., "Compensation of Motion-Induced Phase Errors in TDM MIMO Radars," *IEEE Microwave and Wireless Components Letters*, Vol. 27, No. 12, December 2017, pp. 1164–1166.

[20] Frank, R., "Polyphase Codes with Good Nonperiodic Correlation Properties," *IEEE Transactions on Information Theory*, Vol. 9, No. 1, January 1963, pp. 43–45.

6

Motion and Presence Detection

6.1 Introduction

Detection refers to the process of determining whether or not an echo from a target of interest is present in the radar received signal. Detection has been a primary function of radar systems since the earliest applications of this technology, when radio frequency waves were used to detect the presence of ships [1] and aircraft [2]. Today's radar systems have broadened the scope of detection to include determining the presence of humans, vehicles, other living beings, and inanimate objects.

Detection is made challenging by the presence of other components of the radar received signal, namely noise, clutter, and interference. These processes tend to be nondeterministic, unknown a priori, and potentially varying in their statistical characteristics over time and space.

In this chapter, we will first discuss general detection theory and how it applies to the detection of radar targets in the presence of background noise. We will then present an overview of statistical models of the radar received signal as a combination of target echo and noise, including physical sources and models of target echo fluctuations. Next, we will cover threshold-based detection and the performance of this technique in terms of false positives and probability of detection. We will also present another class of detection techniques known as constant false alarm rate detection and discuss the usage, advantages, and disadvantages compared to absolute thresholding.

In the second part of this chapter, we will discuss the concept of clutter, or unwanted radar echoes. Distinguishing clutter from a signal of interest is an essential function of radar systems. We will cover the impact of clutter on detection and some strategies for mitigating clutter. Finally, we will discuss how the overall design of the signal processing pipeline impacts detection performance.

6.2 Detection Theory

Detection theory originated during World War II as a statistical framework for radar detection of aircraft [3]. Today, modern consumer radar systems are used to detect a wide range of targets, including humans, animals, and consumer vehicles. It is important to appreciate that while the targets of interest may vary greatly, the fundamental principles of determining the presence of a signal of interest amid uncertainty from noise and other sources remain the same. These principles are described by classical detection theory.

Radar-based detection is made challenging by the presence of interference, clutter, and noise, which are typically unknown a priori to the radar system, random in nature, and/or nondeterministically varying [4]. In addition, the characteristics of the target to be detected, such as its radar cross section, motion, position, and orientation relative to the radar, are usually also not precisely modeled, making the signal parameters of received target echo unknown a priori. The many unknown parameters of the signal, clutter, and interference, as well as inherent randomness of noise, combined with the direct impact of detection performance on the radar system's application value, make the field of detection theory a critical component for radar system design.

6.2.1 Hypothesis Testing and Decision Rules

Hypothesis testing provides a mathematical framework in which to formulate the detection problem. In a hypothesis test, each hypothesis H represents a scenario whereby the observed data follows a certain model. The goal of a hypothesis test is to determine, based on the observed data, which hypothesized model is correct. For basic detection, we can formulate a simple binary hypothesis test. The *null hypothesis*, denoted $H = H_0$, represents the scenario that there is no target of interest present at the range, angle, and Doppler coordinates under test. The *positive hypothesis*, denoted $H = H_1$, represents the scenario that the target of interest is present at the range, angle, and Doppler coordinates under test.

The data $(x_1, x_2, ..., x_N)$ is modeled as a random vector with probability density function (PDF) that is conditioned on the true hypothesis. These conditional PDFs are denoted $P_{x|H}(x_1, x_2, ..., x_N|H = H_0)$ and $P_{x|H}(x_1, x_2, ..., x_N|H = H_1)$ for the null and positive hypothesis scenarios, respectively. In Section 6.3, we will delve deeper into specific forms of these PDFs.

A solution to the hypothesis test is a decision rule $\hat{H}(x_1, x_2, ..., x_N)$ that maps all possible instantiations of the observation vector $(x_1, x_2, ..., x_N)$ to a unique hypothesis in the set $\{H_0, H_1\}$ [5]:

$$\hat{H}:(x_1,x_2,...,x_N)\rightarrow\{H_0,H_1\} \tag{6.1}$$

A decision rule thus partitions the space of possible observations into regions corresponding to certain decisions. This partitioning is graphically represented in Figure 6.1 for $N = 2$ and $x_1 \in [010]$, $x_2 \in [010]$.

Since the data is generated by a random process, each decision $\hat{H}(x_1, x_2, ..., x_N) = H_0$ or $\hat{H}(x_1, x_2, ..., x_N) = H_1$ has an associated probability of occurring. The probabilities of making a correct decision or a decision error are denoted by $Pr(\hat{H} = H_i|H = H_i)$ and $Pr(\hat{H} = H_j|H = H_i)$, $i \neq j$, respectively. For the binary hypothesis case, there are four possible combinations of true hypotheses and decisions, described in Table 6.1.

Since

$$Pr(\hat{H} = H_1 \mid H = H_0) = 1 - Pr(\hat{H} = H_0 \mid H = H_0) \tag{6.2}$$

and

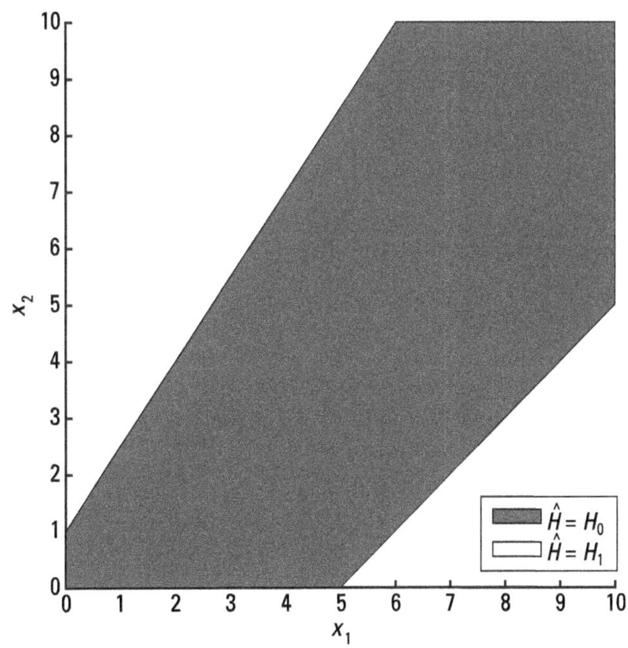

Figure 6.1 Decision regions for an example decision rule for N = 2.

Table 6.1
Possible Outcomes of the Detection Problem, Formulated as a Binary Hypothesis Case

		Decision	
		H_0	H_1
True Hypothesis	H_0	True miss or true negative: target is not present and not detected. Associated cost $C(H_0, H_0)$	False alarm or false positive: target is not present but is detected. Associated cost $C(H_0, H_1)$
	H_1	Missed detection or false negative: target is present but not detected. Associated cost $C(H_1, H_0)$	True detection or true positive: target is present and detected. Associated cost $C(H_1, H_1)$

$$Pr(\hat{H} = H_1 \mid H = H_1) = 1 - Pr(\hat{H} = H_0 \mid H = H_1) \tag{6.3}$$

the probabilistic performance of any decision rule is fully specified by the probability of detection $P_D = Pr(\hat{H} = H_1|H = H_1)$ and probability of false alarm $P_{FA} = Pr(\hat{H} = H_1|H = H_0)$. These probabilities are respectively given in terms of the conditional PDFs and decision regions by

$$P_D = \int_{(x_1,x_2,\ldots,x_N):\hat{H}=H_1} p_X(x_1,x_2,\ldots,x_N \mid H = H_1)\,dx_1 dx_2 \ldots dx_N \tag{6.4}$$

$$P_{FA} = \int_{(x_1,x_2,\ldots,x_N):\hat{H}=H_1} p_X(x_1,x_2,\ldots,x_N \mid H = H_0)\,dx_1 dx_2 \ldots dx_N \tag{6.5}$$

The optimal decision rule $\hat{H}_{opt}(x_1, x_2, \ldots, x_N)$ is the one that minimizes the expected value of an objective function $C(H_i, H_j)$ representing the cost of deciding H_j when the correct hypothesis is H_i. Different cost functions will produce different optimal decision rules, and the exact cost values $C(H_i, H_j)$ are highly specific to the application. Generally speaking, the highest costs result from decision errors (i.e., missed detections and false positives), but the cost of one type of error may be very different from the other. For example, for a gesture control radar system, a false positive may produce a much worse user experience due to the device behaving erratically, compared to a missed detection, where the user might naturally simply repeat the gesture a second time without great detriment of their interaction experience. Conversely, for a security application, a false positive may trigger security personnel to check a camera feed with very low expended effort and cost, while a missed detection could represent a total failure of the system to detect an intruder and potentially lead to great loss of material goods. There may also be costs associated with nonerroneous detection decisions. For radar-based presence sensors in low-power consumer devices, a target detection may trigger a higher power system state such as the device waking from sleep mode. Thus, even a true detection will have a nontrivial cost in terms of power consumption and battery life.

In theory, for a system in which the cost function $C(H_i, H_j)$ is well defined, the optimal decision rule $\hat{H}_{opt}(x_1, x_2, \ldots, x_N)$ can be derived by minimizing the expected cost over all possible decision rules $\hat{H}(x_1, x_2, \ldots, x_N)$:

$$\hat{H}_{opt}(x_1, x_2, \ldots, x_N) = min_{\hat{H}(.)} E\left[C\left(H_i, \hat{H}(x_1, x_2, \ldots, x_N)\right)\right] \tag{6.6}$$

In practice, however, it is rare that the cost function $C(H_i, H_j)$ is perfectly defined. For radar-based interaction systems, for example, quantifying the cost of negative user sentiment caused by detection errors may not be straightforward. Hence, radar systems often use an alternative optimization criterion known as the Neyman-Pearson criterion, described in the next section.

6.2.2 Neyman-Pearson Criterion and Likelihood Ratio Test

The Neyman-Pearson criterion is a common optimization criterion that seeks to maximize the probability of detection, given the constraint that the probability of false alarm cannot exceed a fixed acceptable value P'_{FA}.

The optimal decision rule under the Neyman-Pearson criterion turns out to be a *likelihood ratio test* (LRT) of the following form [6]:

$$\begin{aligned} \hat{H}(x_1, x_2, \ldots, x_N) &= H_1 \text{ if } \Lambda(x_1, x_2, \ldots, x_N) > T \\ \hat{H}(x_1, x_2, \ldots, x_N) &= H_0 \text{ if } \Lambda(x_1, x_2, \ldots, x_N) < T \end{aligned} \tag{6.7}$$

where $\Lambda(x_1, x_2, \ldots, x_N)$ is the ratio of conditional PDFs

$$\Lambda(x_1, x_2, \ldots, x_N) = \frac{p_{X|H}(x_1, x_2, \ldots, x_N \mid H = H_1)}{p_{X|H}(x_1, x_2, \ldots, x_N \mid H = H_0)} \tag{6.8}$$

and T is a threshold value set such that $P_{FA} \leq P'_{FA}$. This likelihood ratio test indicates that, given an instantiation of the data $(x_1, x_2, \ldots, x_N)$, the detection decision depends solely on the statistic $\Lambda(x_1, x_2, \ldots, x_N)$. If $\Lambda(x_1, x_2, \ldots, x_N)$ is greater than a certain threshold T, then the decision should be the positive hypothesis and a detection should be declared. Otherwise, the decision should be the null hypothesis, and no detection should be declared.

Often it is difficult or impractical for a radar system to compute the likelihood ratio $\Lambda(x_1, x_2, \ldots, x_N)$. In many cases, the likelihood ratio can be simplified to an equivalent statistic with a corresponding threshold that is more practical to compute in a real system. We will provide examples of such detection threshold statistics for some common radar signal and signal-plus-noise PDFs in the ensuing subsections.

6.3 Signal and Noise Models

The received radar signal is typically modeled as a sum of contributions from the target of interest, noise, clutter, and interference. In this section, we will first consider probabilistic models of the signal of interest (i.e., the target echo), and noise, temporarily ignoring clutter and interference in order to present the fundamentals of detection theory. In the later portion of this chapter, we will consider how clutter and interference affect detection performance and how their effects can be mitigated.

Let us denote the received signal at a particular range, angle, and/or velocity coordinate by the pulse time-indexed variable $x(T)$. At each pulse time instance T, the received signal $x(T)$ can be modeled as one of the following combinations of a target echo signal at the receiver, $s(T)$, and noise at the receiver, $n(T)$, depending on whether or not a target is present at the coordinates under consideration:

$$\begin{aligned} x(T) &= s(T) + n(T) \text{ if the target is present} \\ x(T) &= n(T) \text{ if the target is not present} \end{aligned} \tag{6.9}$$

In this simplified model, the PDF of the received radar signal $x(T)$ is fully specified by the PDFs of the target echo signal $s(T)$ and the noise process $n(T)$, which are described later.

6.3.1 Target RCS Fluctuations

Random-like fluctuations in the target echo signal $s(T)$ may arise from multiple physical sources. Atmospheric conditions and associated propagation loss may change from one radar transmission to the next. Small variations in the radar system parameters (e.g., the transmit power or frequency) can also cause changes in the received target echo signal. One of the dominant sources of target echo signal variation is due to the fluctuations in the target's radar cross section, σ.

Recall that the radar cross section (RCS) is a measure of the ratio of transmit power intercepted by a target and power reflected by the target back toward the radar. It thus is an abstraction of the scattering characteristics and reflectivity of the target. The received radar signal power is linearly proportional to the target RCS as described by the radar equation. Hence, we can understand the stochastic nature of the received target echo signal by first modeling the randomness in the RCS σ [7].

The radar cross section of a target depends on several characteristics of the target itself as well as the radar system parameters. Small changes in the target position or orientation relative to the radar can cause a change or fluctuation in its RCS and hence received signal strength. The physical origin of these fluctuations is due to multiple scatterers within the target arriving at the radar in phase, out of phase, or somewhere in between, depending on the radar-target geometry. The radar cross section of a hand, for example, can fluctuate as much as 15 dB with changes in the orientation or pose relative to the radar [8].

It is usually intractable to capture the complexity of the target scattering physics and the sensitivity of the RCS on aspect angle and radar frequency in a deterministic RCS model. Moreover, in radar applications, the target scattering physics and aspect angle are usually unknown a priori. Hence, we statistically model the RCS as a random-like parameter with a probability density function.

6.3.1.1 Nonfluctuating

The RCS of an ideal nonfluctuating target can be modeled as a variable with constant but unknown value σ_a. The PDF of the RCS is then a Dirac delta function:

$$p_\sigma(\sigma) = \delta(\sigma - \sigma_a) \tag{6.10}$$

6.3.1.2 Exponential

If we consider the sum of a large number of randomly distributed scatterers, each with approximately the same RCS, then the total received echo from the target will tend to have complex normal Gaussian distribution according to central limit theorem; that is, the real and imaginary components will be independent and identically distributed zero-mean Gaussians:

$$p_{S_R}(s_R) = \frac{1}{\sqrt{2\pi\sigma_a}} e^{-s_R^2/2\sigma_a} \tag{6.11}$$

$$p_{S_I}(s_I) = \frac{1}{\sqrt{2\pi\sigma_a}} e^{-s_I^2/2\sigma_a} \tag{6.12}$$

where σ_a is the variance.

The amplitude of the echo is then Rayleigh distributed, while the power (squared magnitude) and RCS are exponentially distributed.

$$p_\sigma(\sigma) = \frac{1}{\sigma_a} e^{-\sigma/\sigma_a} \tag{6.13}$$

6.3.1.3 Chi-Square

For some targets, the scattering behavior is more realistically modeled as the sum of a single dominant scatterer among a number of lesser scatterers, all randomly distributed. In this case, the RCS is distributed as

$$p_\sigma(\sigma) = \frac{4\sigma}{\sigma_a^2} e^{-2\sigma/\sigma_a} \tag{6.14}$$

Example PDFs of nonfluctuating, exponential, and Chi-square are shown in Figure 6.2 for $\sigma_a = 10$.

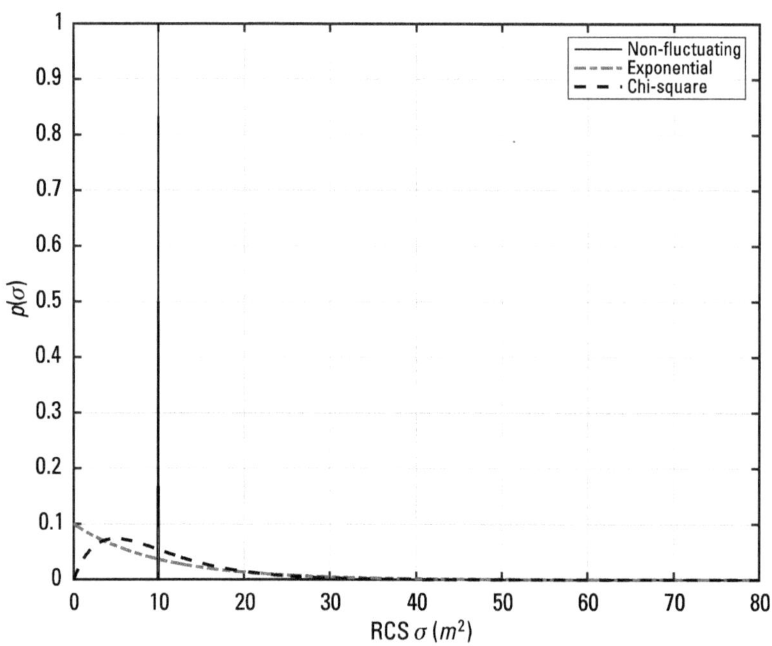

Figure 6.2 Radar cross section PDFs for $\sigma_a = 10$.

6.3.1.4 Other Distributions

There are several other PDFs of radar cross section described in the literature, including the Weibull and log-normal distributions. For more information, we refer the reader to [9, 10].

6.3.1.5 Temporal Decorrelation

The correlation of the observed target RCS from one time instance to the next is dependent on temporal changes in the target, radar system, and their relative geometry and motion. It is illustrative to consider two limiting cases: one in which the observed RCS stays constant over the coherent integration interval (CIT) but is independent from one CIT to the next, and one in which the observed RCS at every pulse is an independent sample from the PDF.

Swerling models are standard descriptions of radar echo statistical behavior [7]. Each model describes a combination of an RCS fluctuation PDF and temporal decorrelation period.

6.3.2 Noise

Noise refers to random contributions to the signal that contain no information about the signal of interest. These contributions can be described as a random process where the samples are mutually uncorrelated as well as uncorrelated with the target echoes.

Typically, the noise in a radar system is modeled as independent, identically distributed standard complex Gaussian noise with zero mean and variance $a^2/2$. The I and Q channels are thus independent and have the following PDFs:

$$p_{N_R}\left(n_R\right) = \frac{1}{\sqrt{\pi a^2}} e^{-n_R^2/a^2} \tag{6.15}$$

$$p_{N_I}\left(n_I\right) = \frac{1}{\sqrt{\pi a^2}} e^{-n_I^2/a^2} \tag{6.16}$$

The magnitude of the noise then has Rayleigh distribution

$$p_N\left(|n|\right) = \frac{2|n|}{a^2} e^{-|n|^2/a^2} \tag{6.17}$$

and the phase is uniformly distributed between 0 and 2π. Examples of complex noise are shown in Figure 6.3. The corresponding magnitude distribution is shown in Figure 6.4.

6.4 Threshold Detection

In this section, we will apply the principles of detection theory discussed in Section 6.2 to the radar signal and noise models presented in Section 6.3. We will show how the Neyman-Pearson likelihood ratio test results in an optimal detection strategy based on thresholding.

For simplicity, we will consider the case where detection is performed on the received signal from a single pulse, which we will denote x. The size of the observation vector is $N = 1$.

6.4.1 Optimal Detection of Nonfluctuating Target

Using the noise model described in Section 6.3.2, the conditional PDF of the received signal given the null hypothesis (target not present) is simply the PDF of the noise, the complex Gaussian with variance denoted by $a_N^2/2$:

$$p_{X|H}(x \mid H = H_0) = \frac{1}{\pi a_N^2} e^{-|x|^2 / a_N^2} \tag{6.18}$$

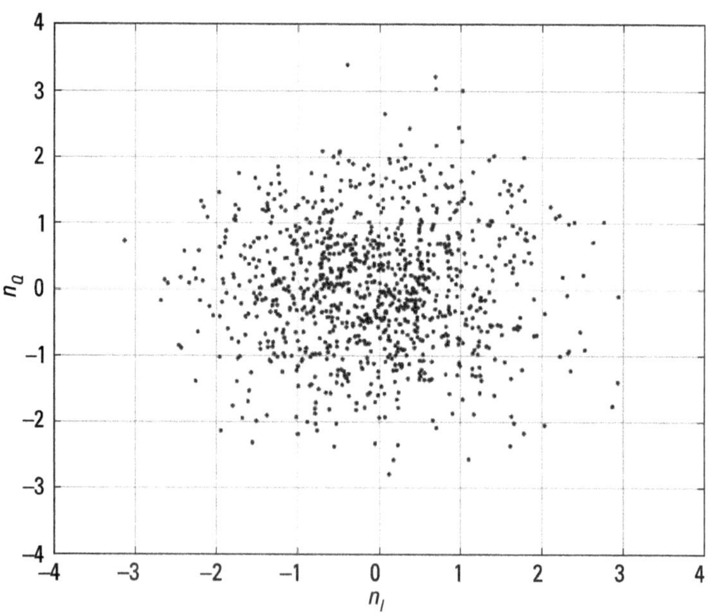

Figure 6.3 Samples of complex Gaussian noise in the I/Q space.

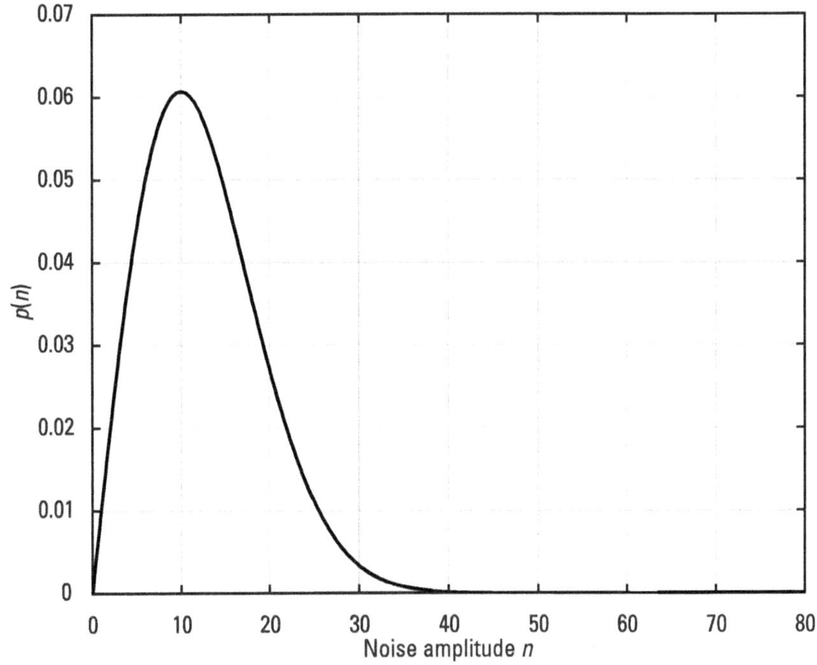

Figure 6.4 Rayleigh distribution describing the PDF of the amplitude of complex Gaussian noise.

It can be shown that when the target echo has constant (but generally unknown) magnitude a_T and the received signal is the sum of the target echo and noise, then the received signal has the Ricean PDF:

$$p_{X|H}(x \mid H = H_1) = \frac{1}{\pi a_N^2} e^{-\left(|x|^2 + a_T^2\right)/a_N^2} I_0\left(\frac{2a_T |x|}{a_N^2}\right) \tag{6.19}$$

where $I_0(.)$is the modified Bessel function of the first kind.

Substituting $p_{X|H}(x|H = H_0)$ and $p_{X|H}(x|H = H_1)$ into the likelihood ratio test in (6.7), is given by

$$\begin{aligned} \hat{H} &= H_1 \text{ if } \Lambda(x) > T \\ \hat{H} &= H_0 \text{ if } \Lambda(x) < T \end{aligned} \tag{6.20}$$

where the likelihood ratio simplifies to

$$\Lambda(x) = \frac{p_{X|H}(x \mid H = H_1)}{p_{X|H}(x \mid H = H_0)} = e^{-a_T^2/a_N^2} I_0\left(\frac{2a_T |x|}{a_N^2}\right) \tag{6.21}$$

Because the right side of (6.21) is a monotonic function of $|x|$, we can equivalently compare $|x|$ alone to a modified threshold:

$$\hat{H} = H_1 \ \text{if} \ |x| > T' \\ \hat{H} = H_0 \ \text{if} \ |x| < T' \tag{6.22}$$

The modified threshold T' is set to satisfy the Neyman-Pearson criterion of constraining the probability of false alarm to a specified level P'_{FA}. To determine the value of T', we evaluate the probability of false alarm as a function of the threshold. As discussed in Section 6.3.2, the amplitude of the complex Gaussian noise has Rayleigh distribution:

$$P_{FA}(T') = \int_{T'}^{\infty} p_{|X|}(|x| \mid H = H_0)\, dx = \int_{T'}^{\infty} \frac{2|x|}{a_N^2} e^{-|x|^2/a_N^2}\, dx = e^{-T'^2/a_N^2} \tag{6.23}$$

Hence for a desired probability of false alarm P'_{FA}, the threshold should be set to

$$T' = \sqrt{-a_N^2 \ln(P'_{FA})} \tag{6.24}$$

The results specify the optimal detector for a nonfluctuating target in complex Gaussian noise using the Neyman-Pearson criterion: the magnitude of the received signal is simply compared to a threshold, which is determined by the maximum acceptable probability of false alarm. If the magnitude of the received signal exceeds the threshold, then a target detection should be declared. Otherwise, no target is detected.

6.4.2 Detection Performance

The performance of this detection strategy can be quantified with the probabilities of false alarm (6.23) and detection as a function of the threshold T':

$$P_D(T') = \int_{T'}^{\infty} p_{|X|}(|x| \mid H = H_1)\, dx = Q\left(\sqrt{\frac{2a_T^2}{a_N^2}}, \sqrt{\frac{2T'^2}{a_N^2}}\right) \tag{6.25}$$

where $Q(.,.)$ is the Marcum Q-function.

Figure 6.5 shows the conditional PDFs of the detection statistic $|x|$ for the null and positive hypothesis, with the probabilities of false alarm and detection denoted by the areas marked in dark grey and grey, respectively, for an example threshold $T' = 3.5$. Figure 6.6 illustrates how the probabilities of detection and

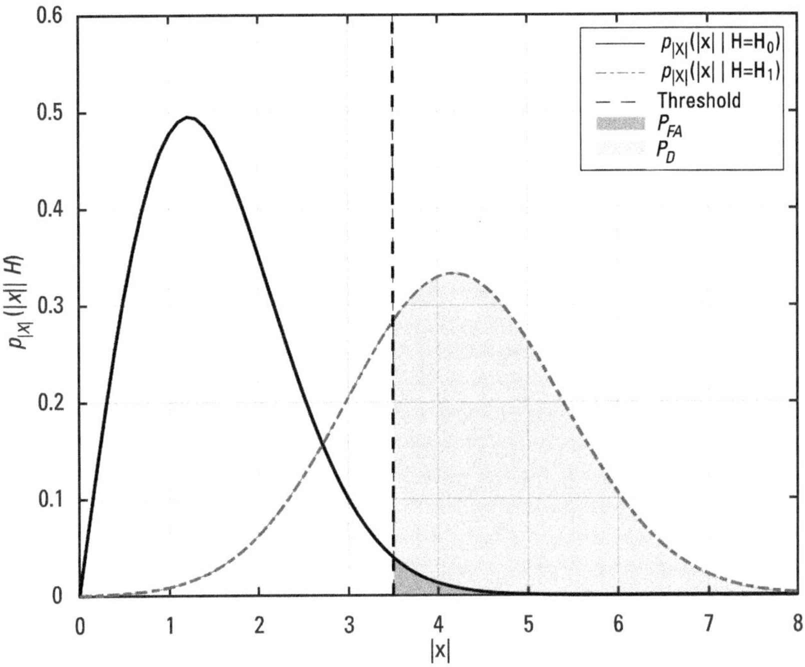

Figure 6.5 Conditional PDFs of the detection statistic $|x|$ for a nonfluctuating target in complex Gaussian noise. An example threshold of $T' = 3.5$ is plotted with the dashed line. The PDFs of the null and positive hypotheses are shown in solid and dotted trace, respectively, with the probabilities of false positive and detection denoted by the areas marked in dark grey and grey, respectively.

false alarm (and hence missed detection and true miss as well) vary as a function of the threshold T' for a nonfluctuating target in complex Gaussian noise.

The tradeoff between probabilities of false alarm and detection can alternatively be represented by a receiver operating characteristic (ROC) curve, shown in Figure 6.7. The point on the ROC curve corresponding to the example threshold of $T' = 3.5$ is indicated by the little square. As the threshold value is decreased, the detection performance moves along the ROC curve in the direction of the arrow. We see that decreasing the threshold results in a higher probability of detection, but a corresponding increase in probability of false alarm as well. The ROC curve provides a useful tool for evaluating detection performance against a hypothetical perfect detector that achieves $P_D = 1$, $P_{FA} = 0$ and a totally random guess, which is represented by the diagonal line. Using the ROC curve, the radar system designer may adjust the detection threshold to achieve a desired tradeoff between true detection rate and false positive detection rate for the application.

The ratio of signal (target echo) power to noise power greatly impacts detection performance. Intuitively, as the target echo power increases relative

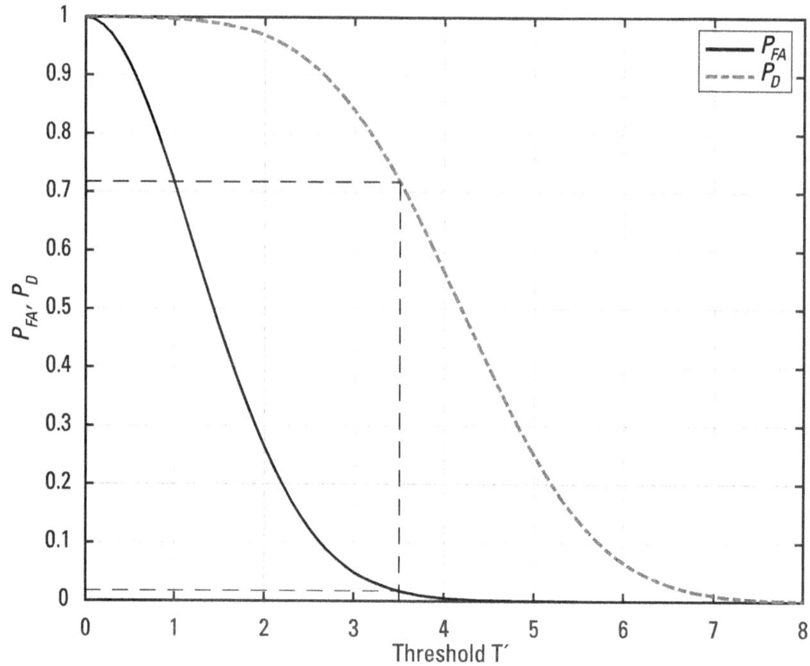

Figure 6.6 The probabilities of false alarm and detection as a function of T' for a nonfluctuating target in complex Gaussian noise. The point on the curves corresponding to the example threshold of $T' = 3.5$ is plotted with the dashed lines.

to the average noise power, the magnitude of a true target echo will become more likely to exceed noise fluctuations in received signal. Hence increasing signal-to-noise ratio results in a fundamental improvement in detection performance; for any value of the probability of false alarm constraint P'_{FA}, a system with higher SNR will have a higher probability of detection than a system with lower signal-to-noise ratio. The impact of SNR on detection performance for the nonfluctuating target case is depicted in Figure 6.8. Indeed, much of radar system and signal processing design is concerned with improving SNR in order to achieve the best possible detection performance.

6.4.3 Impact of Target Fluctuation

The numeric results in Section 6.4 thus far apply to a nonfluctuating target model in complex Gaussian noise. The principles of optimal threshold detection and steps for applying the likelihood ratio test apply for other target and noise models as well (e.g., the more realistic fluctuating target models presented in Section 6.3.1). Deriving the detection statistics, optimal threshold tests, and detection performance for each of these models is beyond the scope of this chapter. Instead, we refer the reader to [11] for more information.

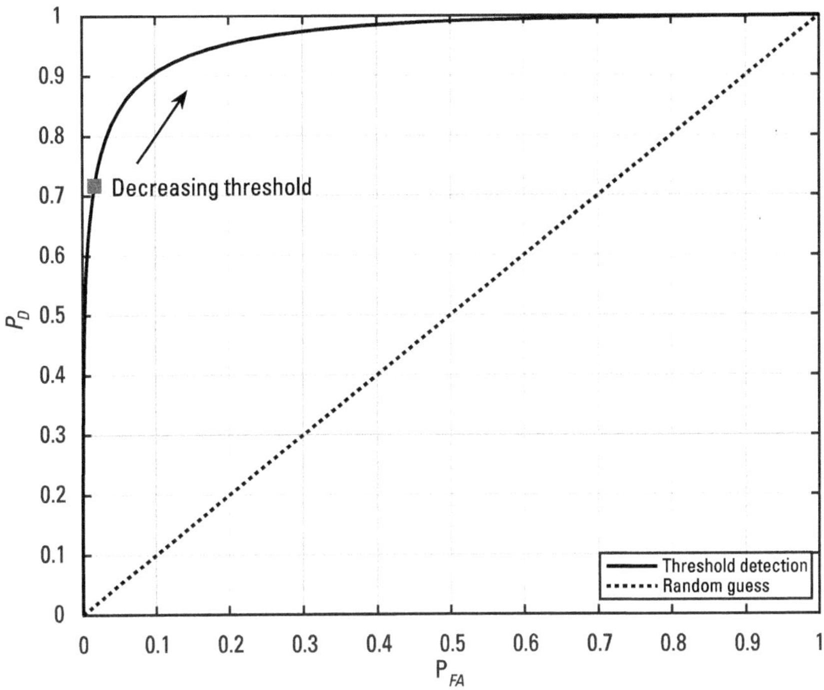

Figure 6.7 Receiver operating characteristic curve for the nonfluctuating target in complex Gaussian noise. The point on the curve corresponding to the example threshold of $T' = 3.5$ is plotted with the square.

6.5 Constant False Alarm Rate Detection

We have shown that when the signal and noise PDFs are well modeled, then the optimal detection strategy is a likelihood ratio test with a fixed threshold determined by the acceptable probability of false alarm. However, many real scenarios do not meet the conditions for optimal fixed threshold detection. Often, noise statistics may be unknown, difficult to model, or vary in time or space. The transceiver gains and other radar parameters affecting absolute power levels may differ slightly from device to device, making it difficult to ensure consistent detection performance across devices with a common fixed threshold approach.

In order to account for unknown or varying noise and absolute signal power levels, we can use a different detection approach called constant false alarm rate (CFAR) detection. As the name suggests, CFAR detection is an adaptive thresholding approach that adjusts the threshold value at each range, angle, and/or Doppler coordinate in order to maintain a constant probability of false alarm. The adjustment of the detection threshold at each coordinate is based on an estimate of the local noise statistics, calculated from data in adjacent or near-adjacent range, angle, and/or Doppler bins.

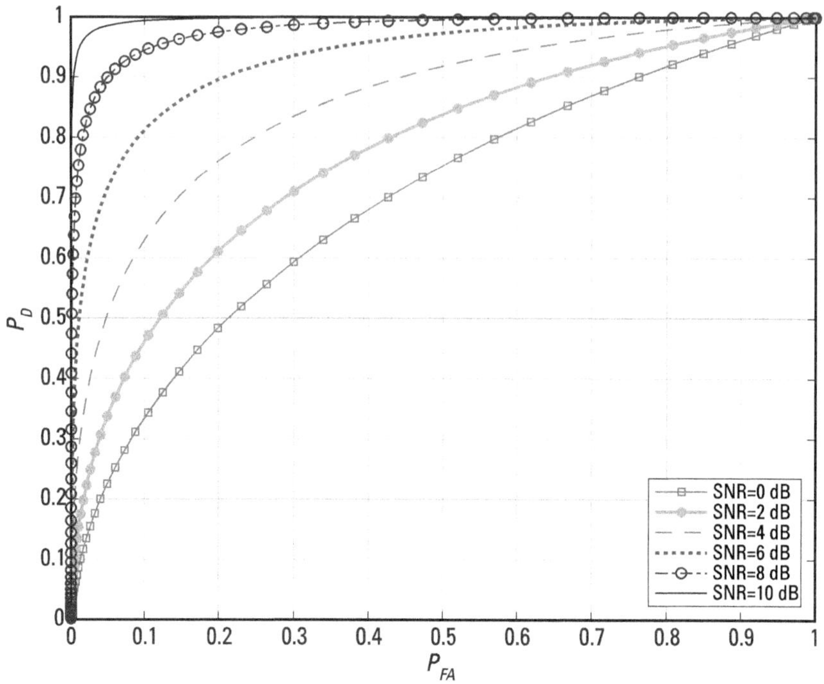

Figure 6.8 ROC curves for nonfluctuating target in complex Gaussian noise for varying signal-to-noise ratios.

The CFAR detection threshold at a certain range, angle, and/or Doppler bin (known as the cell under test) is computed as

$$T = \alpha \hat{g}_n \tag{6.26}$$

where α is a scaling factor dependent on the desired probability of false alarm, and $\hat{g}_n$ is a statistic of the local noise. The statistic $\hat{g}_n$ is computed over a set of adjacent or near-adjacent cells called test (reference) cells, denoted by S_{test}. Typically, these cells are those surrounding the cell under test, excluding a set of guard cells immediately adjacent to the cell under test. The exclusion of guard cells from the noise statistic calculation is intended to prevent leakage from the cell under test from affecting the noise estimate. Figure 6.9 shows an example of how the test cells and guard cells are structured around the cell under test for two-dimensional CFAR in the range-doppler domain.

For a linear detector, the magnitude of the cell under test, denoted $|x_{CUT}|$, is then compared to the threshold to determine which detection hypothesis $\hat{H}$ to declare:

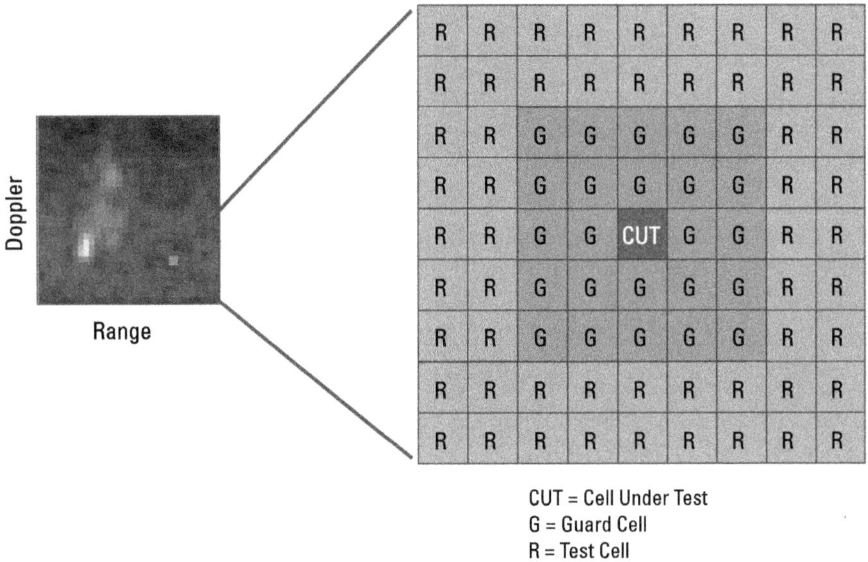

Figure 6.9 Diagram of 2-D CFAR guard and test cells in the range-Doppler domain.

$$\begin{aligned}\hat{H} &= H_1 \text{ if } |x_{CUT}| > T \\ \hat{H} &= H_0 \text{ if } |x_{CUT}| < T\end{aligned} \tag{6.27}$$

Similarly, for a square law detector, the squared magnitude of the cell under test is compared to the threshold.

Several flavors of CFAR techniques are designed for different clutter and noise scenarios, and they vary in the statistic $\hat{g}_n$, the local region over which $\hat{g}_n$ is computed, and how $\hat{g}_n$ is calculated. Next, we will describe a few of the more common CFAR techniques.

6.5.1 Cell-Averaging CFAR

In the cell-averaging CFAR approach, the sample mean of the squared magnitude of the test cells is used as the local noise statistic:

$$\hat{g}_n = \frac{1}{N_{test}} \sum_{i \in S_{test}} |x_i|^2 \tag{6.28}$$

where $N_{test} = |S_{test}|$ denotes the number of test cells.

The threshold scaling factor α to achieve a desired probability of false alarm is given by

$$\alpha = N_{test} \; [(P'_{FA})^{-1/N_{test}} - 1] \tag{6.29}$$

The cell-averaging CFAR technique provides the maximum likelihood estimate of the noise power a_N^2 when the detector is square law; the noise has Rayleigh distributed amplitude with unknown average noise power; and the noise is independent and identically distributed across the cell under test and noise test cells. It can be shown [12] that the expected probabilities of false alarm and detection are then

$$P_D = \left[1 + \frac{\alpha / N_{test}}{1 + SNR}\right]^{-N_{test}} \tag{6.30}$$

$$P_{FA} = \left[1 + \frac{\alpha}{N_{test}}\right]^{-N_{test}} \tag{6.31}$$

As intended, the probability of false alarm is constant across all values of noise.

The cell averaging CFAR technique is sufficient for simple homogeneous interference when targets are unlikely to be closely spaced to each other and to clutter. However, it is susceptible to target masking if another target or clutter happens to be located in the test cells. In this situation, the computed value of $\hat{g}_n$ is erroneously biased by the echo of the secondary target, causing the CFAR threshold to exceed the magnitude of the cell under test even with the presence of the primary target echo.

Target masking can be partially mitigated by employing more robust CFAR techniques, such as the greatest-of modification described next.

6.5.2 Greatest-of and Least-of CFAR

Some CFAR techniques divide up the set of test cells into subsets: For example, one subset of test cells with doppler frequencies greater than the cell under test and the other subset with doppler frequencies less than the cell under test. The noise statistics are then computed over each subset separately. In the greatest-of (GO) CFAR technique, the subset producing the largest estimate of noise power is used to determine the threshold for the cell under test. In GO CFAR, the local noise statistic is obtained as:

$$\hat{g}_n = \text{the greatest of} \frac{1}{N_{test1}} \sum_{i \in S_{test1}} |x_i|^2 \text{ and} \frac{1}{N_{test2}} \sum_{i \in S_{test2}} |x_i|^2 \tag{6.32}$$

where S_{test1} is the first subset of the test cells, and S_{test12} is the second subset of the test cells.

Conversely, in the least-of (LO) CFAR technique, the subset producing the smallest estimate of noise power is used:

$$\hat{g}_n = \text{the smallest of} \frac{1}{N_{test1}} \sum_{i \in S_{test1}} |x_i|^2 \text{ and} \frac{1}{N_{test2}} \sum_{i \in S_{test2}} |x_i|^2 \tag{6.33}$$

The GO and LO techniques are valuable to consider for human motion sensing applications due to their ability to deal with clutter without greatly increasing computational complexity over simple cell averaging CFAR. For example, in gesture recognition applications, a least-of cell averaging CFAR technique can mitigate clutter reflections from the torso and head, preventing target masking of the hand.

6.5.3 Ordered Statistics CFAR

In ordered statistics (OS) CFAR [13], the test cells are ordered by decreasing amplitude:

$$|x_{(1)}| > |x_{(2)}| > \cdots > |x_{(N_{test})}| \tag{6.34}$$

where the indexes in the parentheses denote the rank order. The noise estimate is then computed over the k_{th} element of the ordered cells. OS CFAR is useful for dealing with noise characteristics that tend to have large outliers or multiple targets.

6.6 Clutter Rejection

An important component of detection is rejecting other components of the signal that are not the signal of interest.

In general, we use the term *clutter* to refer to received reflections from objects other than the target of interest. It is important to understand that the definition of clutter therefore varies with the purpose of the radar system. For a gesture sensing radar system, clutter may include returns from furniture, device enclosure, or body parts other than the hand. For an environment or object recognition system, humans and other live beings may be considered clutter.

An important distinction between clutter and noise is that, like the signal of interest, clutter is an echo of the transmitted signal. Hence, many system design choices to improve signal power will also increase the clutter power (e.g,. increasing transmit power). The ratio of signal power to clutter power is often characterized by the signal-to-clutter ratio. A key figure of merit for clutter mitigation techniques is the improvement factor, or change in signal-to-clutter ratio. In this section, we will discuss techniques to increase the signal-to-clutter ratio for improved detection of the target of interest.

In traditional signal processing approaches, clutter attenuation techniques are designed or decided based on differentiating characteristics of the target of interest and clutter. These require prior understanding and models of the target and clutter. Clutter may be differentiated from the target of interest based on its location or motion characteristics. For instance, we may be interested in targets within a certain distance of the radar and moving with some minimum velocity.

Today, there is growing interest and potential in using machine learning to automatically learn the differentiating characteristics of the target of interest and clutter based on data sets. In this section, we will cover some existing approaches to clutter reduction to build intuition.

6.6.1 Regions of Interest

Detection is generally performed across multiple bins of the signal domain in a region of interest. The specification of the region of interest is one approach to eliminating clutter.

6.6.2 Doppler Filtering

A common type of clutter for presence and motion detection is environmental clutter, or radar returns from stationary objects in the environment (e.g., furniture and walls). These can be differentiated from human targets based on their appearance in the Doppler spectrum. In many consumer radar cases, the radar is assumed to be stationary. In those cases, stationary environmental clutter will appear at zero Doppler.

Various filter designs may be applied in the slow time domain to attenuate reflections with a certain Doppler frequency. The field of filter design is beyond the scope of this chapter. Instead, we will present a few simple digital filters that can be used to mitigate stationary clutter. These filters can be cascaded with the usual range-Doppler processing.

6.6.2.1 FIR Filters

A finite impulse response (FIR) filter has an impulse response of finite duration. A simple FIR stationary clutter filter is the N-pulse canceler, which can be implemented with a simple delay line. The two-pulse canceler is given by

$$y(T) = x(T) - x(T-1) \tag{6.35}$$

The frequency response or transfer function of the two-pulse canceler is

$$H(f) = 2je^{-j\pi f T_{pulse}} sin\left(\pi f T_{pulse}\right) \tag{6.36}$$

where T_{pulse} is the pulse repetition interval. The amplitude of the frequency response is plotted in Figure 6.10.

Hence, targets at zero Doppler or multiples of the PRF will be attenuated.

6.6.2.2 IIR Filters

Conversely, an infinite impulse response (IIR) filter has an impulse response of infinite length.

$$c(T) = (1-\beta)x(T-1) + \beta c(T-1) \tag{6.37}$$

$$y(T) = x(T) - c(T) \tag{6.38}$$

where $\beta \leq 1$ is a parameter that tunes the filter response. Note that this filter is recursive; that is, the clutter map $c(T)$ at any time T is dependent on the previous state of the clutter map.

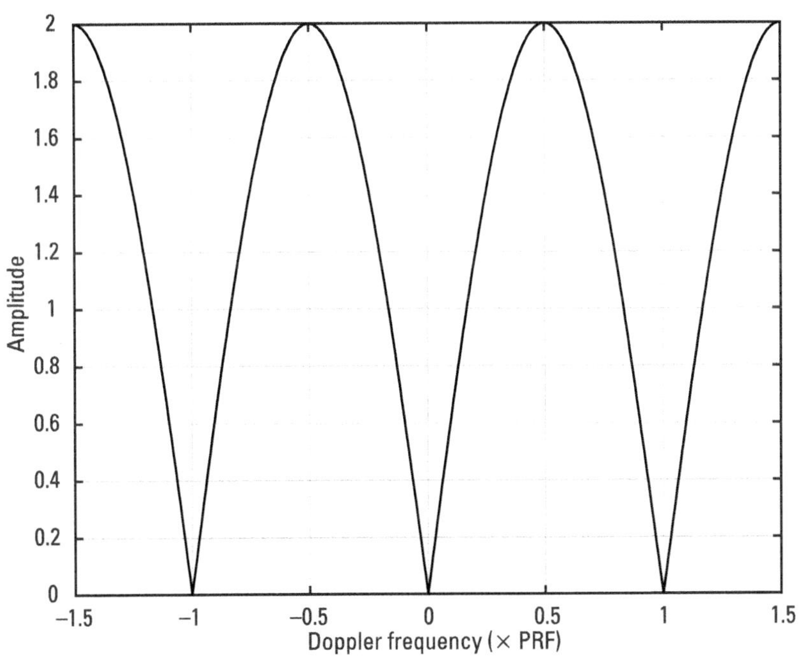

Figure 6.10 Frequency response of two-pulse canceler.

The combined effect of this filter is a high pass filter that attenuates signals near zero frequency. The frequency response of the overall filter is discussed in (4.48) and (4.49). As the filter tap value β increases, the cutoff frequency decreases and the filter notch becomes sharper. A sharper notch is often desirable because it enables the retention of very slow moving targets of interest that are near zero Doppler but not completely stationary.

There is a clear tradeoff between the frequency sharpness of the filter and the rate of decay of the response to perturbations. As β increases, the filter notch becomes sharper, enabling the retention of very slow moving targets of interest; at the same time, any perturbation produces a response that takes longer to die away.

6.6.2.3 Frequency-Domain Filters

Doppler filtering can also be performed in the frequency domain instead of the time domain. This is particularly useful for clutter that has a unique micro-Doppler structure (e.g., audio devices have physical vibration actuators to produce sound waves). These vibrations may cause the device to produce radar echoes with a symmetric Doppler spectrum (an example is shown in Figure 6.11). Hence, a frequency-domain filter can be designed to exploit this symmetry in order to remove these audio vibrations from the radar signal.

6.6.3 Spatial Filtering

Clutter can be attenuated in the spatial domain by applying beam nulling techniques. Such algorithms produce a spatial notch filter that attenuates reflections

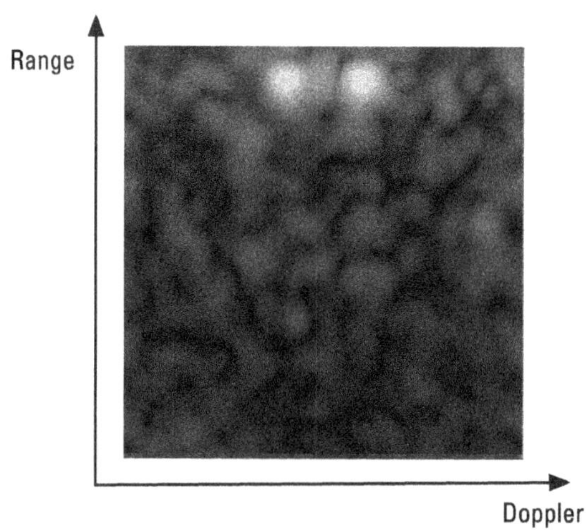

Figure 6.11 Range-doppler signal showing characteristic symmetric Doppler spectrum of vibrating audio components.

arriving from a certain direction. This can be helpful if, for example, the radar has a fixed orientation in a room and is detecting human presence. By nulling reflections arriving from the ceiling or floor, we can potentially remove interfering radar echoes due to ceiling fans or small pets.

Consider a radar with a linear antenna array of two elements with spacing $d = \lambda/2$ as shown in Figure 5.2(a). A simple beam nulling filter to attenuate clutter from a direction θ has the form

$$y = x_1 - x_2 e^{j2\pi d sin(\theta)/\lambda} = x_1 - x_2 e^{j\pi sin(\theta)} \tag{6.39}$$

where x_1 is the output of antenna 1 and x_2 is the output of antenna 2. This is a simple FIR filter implemented in the spatial domain, analogous to the FIR filter implemented in the slow time domain for Doppler filtering.

6.6.4 Adaptive and Machine Learned Clutter Filters

In classical signal processing approaches, we hand-engineer clutter filters based on assumed or known models or characteristics of the clutter; (e.g., its Doppler spectrum or angular position relative to the radar). The filter parameters can also be adaptively set based on the statistics of the data rather than statically predetermined. Adaptive null-steering is commonly used for dealing with radar jamming or interference that may change its location over time or range, as discussed in Chapter 5. A well-known technique based on adaptive filtering is space-time adaptive processing, in which the parameters of a joint spatial-Doppler filter are estimated based on the data.

Similarly, machine learning provides a framework for automatically learning filters based on data sets, rather than hand-engineering them for a priori models of the target and clutter. Chapter 7 provides an overview of machine learning techniques for radar presence and motion detection.

6.7 Interference

Interference generally refers to components of the signal that are not a result of reflections. For example, RF interference or jamming may occur due to other radio frequency transmitters in the vicinity of the radar, including communication devices that have a fundamental or harmonics in the same frequency range. Interference may also result from coupling or noise on the radar board (e.g., the traces or the power supply).

In many cases, interference can be mitigated using similar filters as clutter. For example, beam nulling is a useful technique for reducing interference from other RF transmitters (e.g., communication devices).

6.8 Detection Pipeline Design

The general theory of detection is agnostic to the domain in which the signal is represented. Hence, detection may be performed after any stage of the signal processing pipeline. Indeed, determining the best domain to perform detection and designing the signal processing steps prior to detection are integral components of the radar detection pipeline and directly impact detection performance, as we will discuss next.

A detection pipeline generally consists of the following components:

- Signal transformation: transformation of the signal into a space where signal can be separated from noise, clutter, and interference. Signal-to-noise ratio is increased through a subset of coherent and noncoherent integration steps.
- Filtering: removal of noise, clutter, and/or interference.
- Signal-of-interest detection.

References

[1] Griffiths, H., et al., "Christian Hülsmeyer: Invention and Demonstration of Radar, 1904," *IEEE Aerospace and Electronic Systems Magazine*, Vol. 34, No. 9, 2019, pp. 56–60.

[2] Watson-Watt, R., "Radar in War and in Peace," *Nature* 156, 1945, pp. 319–324.

[3] Marcum, J., "A Statistical Theory of Target Detection by Pulsed Radar," *IRE Transactions on Information Theory*, 1960, pp. 59–267.

[4] Skolnik, M., *Introduction to Radar Systems*, Second Edition, New York: McGraw-Hill, 1980.

[5] Willsky, A. S., et al., "Stochastic Processes, Detection and Estimation," Course Notes for MIT 6.432, 2003, p. 109.

[6] Neyman, J., and E. S. Pearson, "On the Problem of the Most Efficient Tests of Statistical Hypotheses," *Philosophical Transactions of the Royal Society of London*, Series A, Containing Papers of a Mathematical or Physical Character 231.694-706, 1933, pp. 289–337.

[7] Swerling, P., "Probability of Detection for Fluctuating Targets," *IRE Transactions on Information Theory*, 1960, pp. 269–308.

[8] Hügler, P., et al., "RCS Measurements of a Human Hand for Radar-Based Gesture Recognition at E-band," *German Microwave Conference* (GeMiC), 2016, pp. 259–262.

[9] Schleher, D. C., "Radar Detection in Weibull Clutter," *IEEE Transactions on Aerospace and Electronic Systems*, Vol. 12, No. 6, 1976, pp. 736–743.

[10] Pollon, G., "Statistical Parameters for Scattering from Randomly Oriented Arrays, Cylinders, and Plates," *IEEE Transactions on Antennas and Propagation*, Vol. 18, No. 1, 1970, pp. 68–75.

[11] Richards, M. A., et al., *Principles of Modern Radar: Basic Principles*, Raleigh, NC: Scitech Publishing, 2010.

[12] Raghavan, R. S., "Analysis of CA-CFAR Processors for Linear-law Detection," *IEEE Transactions on Aerospace and Electronic Systems*, Vol. 28, No. 3, 1992, pp. 661–665.

[13] Rohling, H., "Ordered Statistic CFAR Technique—An Overview," *12th International Radar Symposium* (IRS), 2011, pp. 631–638.

7

Radar Machine Learning

Nicholas Gillian

In this chapter, we discuss the application of machine learning to radar sensing. At its core, the objective of radar sensing is to extract and detect meaningful patterns from radar signals. The field of machine learning is particularly adept at detecting subtle patterns in data, whether that is classifying data into discrete categories, detecting and localizing targets, spotting outliers and abnormalities, or mapping a continuous input signal to a more useful continuous output signal. Applying machine learning to radar signals that include human motion and gestures is advantageous, given the diversity of signals that can be returned from users. This chapter therefore focuses on how machine learning can be applied successfully to radar.

7.1 Machine Learning Fundamentals

Before discussing the specific application of machine learning to radar, we first review the fundamentals of machine learning for readers new to this field. While we cover the fundamental principles and concepts behind machine learning, this is in no way a comprehensive discussion of the broad field of machine learning. For readers that wish to learn more, we recommend excellent resources such as [1–4].

Machine learning is a branch of artificial intelligence based on the principle that systems can learn to discover patterns in data and then apply this knowledge to make predictions about new, previously unseen, data.

Tom Mitchell [4] provided a formal definition of computational learning as "A computer program is said to learn from experience **E** with respect to some class of tasks T and performance measure P if its performance at tasks in T, as measured by P, improves with experience E."

For example, instead of manually coding an algorithm to accomplish a task, such as detecting a target from a radar signal (T), we could instead provide a machine learning algorithm a series of recordings (E) that do and do not contain targets of interest and have the algorithm learn to detect targets by optimizing the parameters of a model to achieve high target detection accuracy (P) on a set of test recordings.

This is a fundamentally different approach to the development of standard algorithms. With machine learning, the expert designer now plays the role of setting the constraints and objectives of the system and then providing the "best" possible data to one or more abstract machine learning algorithms that will generate a model that results in the actual "algorithm." Selecting a different dataset or formulating the machine learning task as a different problem can result in fundamentally different results. The goal of this chapter is therefore to provide fundamental principles in how machine learning applies to radar to avoid treating radar-based machine learning as alchemy; to instead view it as fundamental science and engineering.

We primarily focus our discussion throughout this chapter on an area of machine learning called *supervised learning*, due to its wide applicability and success in radar and wider machine learning fields [3, 5–8]. There are other subfields within machine learning that are relevant to radar sensing, such as self-supervised learning and meta learning, which we briefly discuss at the end of this chapter.

7.1.1 Supervised Learning

Supervised learning is an area of machine learning where the goal is to learn a function, f, described by the parameters θ, that maps an input, x, to a desired output, y:

$$y = f_\theta(x) \tag{7.1}$$

For example, in a classic target detection task, x could represent the data from a radar sensor; y could represent the values of either {0, 1} that indicates the presence or absence of a target; and the function f_θ could represent a simple linear model that maps $x \rightarrow y$. In this example, θ would represent the parameters of the function, f.

The term supervised learning stems from the use of a dataset, D, consisting of M pairs of $\{x,y\}$ examples, where the output for each given input is explicitly provided before estimating the function parameters θ.

$$D=\{\{x_1,y_1\},\{x_2,y_2\},\ldots,\{x_m,y_m\}\} \tag{7.2}$$

where x is a tensor[1] with size N, and y is a K-dimensional vector. The size and shape of N and K will vary based on the machine learning task.

Supervised learning algorithms exploit the explicit knowledge of what is the expected output for a given input to estimate the function parameters that robustly maps $x \rightarrow y$. This gives the machine learning algorithm the "correct" answer, which the algorithm leverages to guide it toward the optimal function parameters.

Explicitly knowing what the expected output should be for a specific machine learning input is the primary reason why supervised learning has been so successful. This is particularly the case in fields such as computer vision or natural language processing, where large, high-quality, labeled datasets exist such as ImageNet [9]. However, the requirement for high-quality ground truth labels is also the Achilles heel for supervised learning—as large datasets must first be constructed for each new use case or sensor, which can be incredibly expensive and time consuming. This makes it problematic, particularly for sensors such as radar that lack large pre-established datasets. Furthermore, sensors such as radar can be difficult for humans to manually annotate, as it can be challenging for even expert raters to interpret the exact contents of the signal without other ground truth sources such as cameras or motion capture. We discuss potential solutions to this problem later in this chapter. Despite the limitations of supervised learning, it is still a very powerful tool and having a fundamental understanding of the core concepts is a worthwhile investment.

A supervised learning algorithm can be used to estimate the parameters of a function during a *training* phase, resulting in a *model* defined by the specific parameters θ. The model can be used in an *inference* phase to estimate y for new values of x, even if a specific instance of x was not previously present in the original dataset used to train the model.

If the true value of y for a given input is known in advance, then the predictive performance of the model can be measured by computing the difference between the model's estimated output, $\hat{y}$, and the expected output, y. As we will discuss shortly, the difference between y and $\hat{y}$ can be leveraged to help adjust the parameters of the model to minimize the overall error of the model and *learn* optimal parameters.

1. From a high-level data structure perspective, a tensor is a generic multidimensional array. Tensors can represent a [1] dimensional array (a scalar); an [n] dimensional array (a vector); an [n, m] dimensional array (a matrix); or even higher dimensions [n, m, q, . . .]. From a mathematical perspective, tensors are more specific and represent an algebraic object that describes a multilinear relationship between sets of algebraic objects related to a vector space.

7.1.2 Linear Regression

We begin our discussion of supervised learning algorithms with the task of linear regression. While this may be a trivial task, we will see that the fundamental machine learning concepts applied to solve this problem scale to more advanced tasks and algorithms later in this chapter.

7.1.2.1 Linear Regression Toy Problem

To motivate the discussion of linear regression, let's assume that we have a $1 radar sensor from the billion dollar ToyRadarProblems company. The sensor generates $\mathrm{x} = [x_1, x_2]^T \in \Re$: a two-dimensional signal, where x_1 represents secret radar feature one and x_2 represents secret radar feature two. The manufacturer of the radar sensor won't disclose what these secret features are (this is why they are a billion dollar company), but you notice the values appear to correlate with the distance of a user to the sensor (i.e., when the magnitude of x_1 and x_2 are small, a user is close to the sensor; when the magnitude of x_1 and x_2 are large, the user is far from the sensor). Based on this, you want to train a machine learning model to estimate the actual range of a user to the sensor in centimeters. To try this, you record a small dataset of 100 examples of sensor values from the radar alongside the true distance of the user to the device. A subset of the recordings are shown in Figure 7.1.

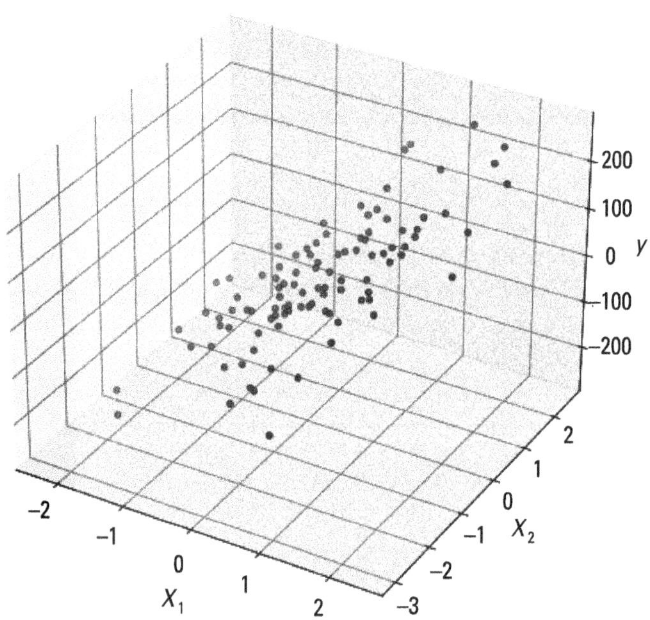

Figure 7.1 Sample recording from the ToyRadarSensor showing the features x1 and x2 and their relationship to y, which represents the user's distance to the sensor.

7.1.2.2 Linear Regression Problem Definition

To perform supervised learning on this task, we must decide how to represent the problem. As an initial choice, let's assume that the problem of mapping from $x \rightarrow y$ can be represented as a linear function of x:

$$y = f_{\theta}(x) = \theta_0 + x_1\theta_1 + x_2\theta_2 \tag{7.3}$$

Here θ are the parameters of the model that parameterize the space of all possible linear functions in $\Re^2$ mapping $x \rightarrow y$. For linear regression, these parameters are more generally referred to as *weights* and denoted by the symbol $w = [w_0, w_1, w_2]^T$; we therefore rewrite (7.3) as:

$$y = f_w(x) = w_0 + x_1w_1 + x_2w_2 \tag{7.4}$$

w_0 is the intercept term and defines where on the y domain the line that maps $x \rightarrow y$ intercepts when $x_1w_1 + x_2w_2 = 0$. To simplify notation, it is common practice to add an additional element to the input vector at x_0 with the constant value of 1 as illustrated in Figure 7.2, giving $x = [1\ x_1\ x_2]^T$.

This additional constant allows us to compute the dot product between x and w. With x_0 set to the constant of 1, it is easy to see the mathematical equivalence of the following representations:

$$w_0 + x_1w_1 + x_2w_2 = x_0w_0 + x_1w_1 + x_2w_2 = \sum\nolimits_{j=0}^{N} x_jw_j = w^T x \tag{7.5}$$

where $N = 2$ is the number of elements in x. Equation (7.4) can therefore be simplified as the dot product between x and w:

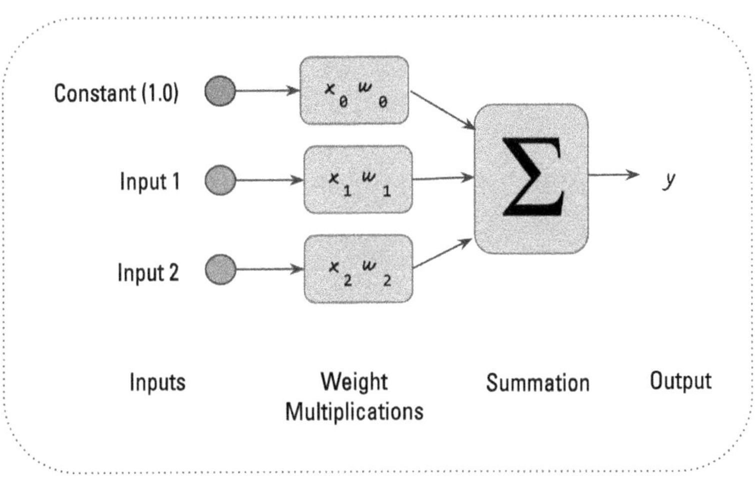

Figure 7.2 A visual representation of a linear regression model with two inputs.

$$y = f_w(x) = w^T x \tag{7.6}$$

7.1.2.3 Linear Regression and Straight Lines

Astute readers will have noticed the relationship between this linear function and the classic equation for a straight line: $y = mx + b$. This is not an accident. Indeed, $w_0 \cong b$, the intercept, and $\{w_1, w_2\} \cong m$, which gives us the slope of the line as shown in Figure 7.3. Understanding this relationship gives us a good intuition of what the weight parameters of the linear model represent, but how can we select the optimal values of w to provide the best estimates for y?

7.1.2.4 Estimating Model Parameters

In a supervised learning context, the optimal values for w can be estimated using the examples of $\{x, y\}$ in the training dataset, where the specific ith pair is denoted $\{x^{(i)}, y^{(i)}\}$. To achieve this, we need to define a function that measures for each potential value of w how close the outputs of $f_w(x^{(i)})$ are to the corresponding $y^{(i)}$. One such distance function or error function is the squared error between the expected output of the mapping function, $\hat{y}^{(\mathrm{i})}$, and the actual output, $y^{(i)}$, for the ith training example:

$$SQUARED\ ERROR(y, \hat{y}) = (y - \hat{y})^2 = (\mathrm{y} - w^T x)^2 \tag{7.7}$$

The motivation for this error function is clear. The squared error between the output of $f_w(x^{(i)})$ and the expected output will be close to zero if the

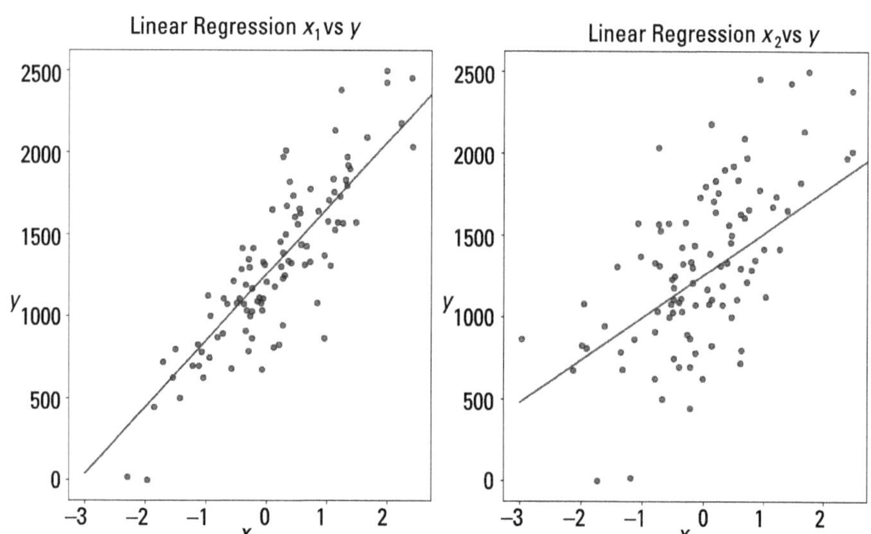

Figure 7.3 Fitting a straight line to the data points for both x1 and x2.

parameters w correctly map $x \rightarrow y$. Alternatively, the squared error will grow exponentially if the distance between the output of $f_w(x^{(i)})$ is far from the expected output $y^{(i)}$. Squaring the error has the nice property of penalizing estimates that are further from the true value, while treating positive or negative distance errors equally.

There are many strategies for how this distance function can be used to estimate the optimal parameters for w given a training dataset. A naive strategy would be to randomly pick many potential values of w, measure the distance for each candidate, and select the candidate with the minimum distance. While this naive strategy might provide an acceptable solution for trivial problems, it should be clear to the reader that this technique is suboptimal at best and does not scale for real-world applications—particularly those where the number of parameters in the model are in the order of millions or billions. So is there a better way to estimate w?

7.1.2.5 Least Mean Squares

Readers with a statistical background will be aware of the well-known closed-form solution for this problem, which is to minimize the sum-of-squares of the errors between the estimates $\hat{y}$ and the targets y over the entire dataset of M examples. This is better known as *ordinary least squares* (OLS) optimization. The OLS algorithm aims to find the parameter weights that minimize the sum of squares error over the entire dataset of M examples:

$$w = arg\,\underset{w}{min} \sum\nolimits_{i=1}^{M} (y^{(i)} - f_w\left(x^{(i)}\right))^2 \tag{7.8}$$

By defining the design matrix X as the M by $N+1$ matrix that contains all Mx training examples, and Y as the M-dimensional vector containing all y training examples, (7.8) can be written in matrix form as:

$$w = arg\,\underset{w}{min}\ (Y - Xw)^T\left(Y - X\mathrm{w}\right) \tag{7.9}$$

Using the OLS algorithm, the unknown parameters w can be solved using the closed-form solution (assuming that the matrix X^TX can be inverted):

$$w = (X^T X)^{-1} X^T Y \tag{7.10}$$

What if there was no closed-form solution to solve the optimal parameters of w for a given dataset? Is there another solution to estimate w? Thankfully, there is. This is where we can combine the core concepts of the OLS algorithm with one of the most important optimization algorithms in machine learning: *gradient descent.*

7.1.2.6 Gradient Descent

Gradient descent is an iterative optimization algorithm that can be used to find the local minimum of any function that is differentiable. This means that if we can define an error function that is differentiable with respect to the weights of a machine learning model, we can use gradient descent to adjust the weights of the model to minimize the error. If we repeat this process over multiple iterations, we can continually reduce the error of the model and thus learn to improve the model, at least on the training dataset.

Gradient descent can be formally defined as:

$$\theta_{t+1} = \theta_t - \eta \Delta f_{\theta_t}(x) \tag{7.11}$$

where θ_{t+1} is the new model parameters at step $t+1$ after the update step, θ_t is the old model parameters used at step t, η is the learning rate, and $\Delta f_{\theta_t}(x)$ is the gradient of the function $f_{\theta_t}(x)$.

A common analogy for gradient descent is to imagine you are a hiker standing on the side of a mountain in thick fog and you wish to reach the safety of the village at the bottom of the mountain. Due to the fog, you can't see in which direction the village is, but all you know is that it is at the bottom of the mountain. Given this knowledge, you use your walking stick to probe the ground to your left, right, front, and back, and can feel in which direction is uphill and which direction is downhill. You can therefore take a step in the downhill direction and repeat the probe with your walking stick. If you repeat this process hundreds of times you can reach the safety of the village, assuming you don't reach a flat plateau or wander into a mountain lake along the way. Gradient descent works in the same way, except instead of manually probing the steepness of the error function at multiple locations around the current position, gradient descent uses the observation that (assuming the function is defined and differentiable) the error will decrease fastest if one goes in the negative gradient of the function, evaluated with respect to the current parameters.

7.1.2.7 Loss (Cost) Functions

To use gradient descent to search for the optimal parameters of our model requires us to define three specific attributes for our problem:

- $J(w)$: a *loss function*; this represents the property we wish to minimize, such as the squared error defined earlier;
- $\frac{\delta}{\delta w_j} J(w)$: the *partial derivatives* of the loss function with regard to the model parameters;
- η: the *learning rate*; this controls how quickly we update the model weights at each iteration.

For our linear regression problem, we will use the squared error as the *loss function.* Instead of minimizing the loss on a single example, we will estimate it over all M training examples. This gives us the following *cost function*:

$$J(w)=\frac{1}{2}\sum_{i=1}^{M}(f_w\left(x^{(i)}\right)-y^{(i)})^2 \tag{7.12}$$

It is worth highlighting two subtle mathematical details in (7.12). These details are added to simplify computing the partial derivatives of $J(w)$, which is required momentarily. First, the constant 1/2 is added to cancel the squared power during differentiation. Second, the order of the terms within the squared operation are carefully placed so that $y^{(i)}$ is subtracted from $f_w(x^{(i)})$. It should be easy for readers to prove that due to the squared operation, $(f_w(x^{(i)}) - y^{(i)})^2$ is identical to $(y^{(i)} - f_w(x^{(i)}))^2$, but this reordering helps simplify the partial derivatives of $J(w)$.

The cost function can now be used to compute the error on the training dataset with respect to the current parameters of the model. To update the weights to reduce the error on the dataset, we can take the partial derivative of the error with respect to each weight and apply a small update to that weight to reduce the error:

$$w_j := w_j - \eta\frac{\delta}{\delta w_j}J(w) \tag{7.13}$$

Here, η is the *learning rate,* which controls the magnitude of change for each weight update. In practice, the learning rate is typically set to a small value (e.g., 0.001) to minimize large jumps of the model throughout weight space.

For the algorithm to work, we need to define the partial derivative for our specific cost function:

$$\frac{\delta}{\delta w_j}J(w) \tag{7.14}$$

To simplify the estimation of $\frac{\delta}{\delta w_j}J(w)$, let's assume for a moment we have just one $\{x,y\}$ example in the training dataset. This lets us temporarily ignore the superscript i and summation in the definition of our cost function, which gives us:

$$J(w)=\frac{1}{2}(f_w(\mathrm{x})-\mathrm{y})^2 \tag{7.15}$$

The partial derivative of $J(w)$ on a single example can therefore be derived as follows:

$$
\begin{aligned}
\frac{\delta}{\delta w_j} J(w) &= \frac{\delta}{\delta w_j}\frac{1}{2}(f_w(x)-y)^2 \\
&= 2\frac{1}{2}(f_w(x)-y)\frac{\delta}{\delta w_j}(f_w(x)-y) \\
&= (f_w(x)-y)\frac{\delta}{\delta w_j}\sum\nolimits_{j=1}^{N} x_j w_j - y \\
&= (f_w(x)-y)x_j
\end{aligned}
\tag{7.16}
$$

For a single training example, the update rule therefore simplifies from (7.13) to:

$$
w_j := w_j - \eta\,(f_w(x)-y)x_j \tag{7.17}
$$

Expanding this back to all M examples in the training dataset gives:

$$
w_j := w_j - \eta\sum\nolimits_{i=1}^{M}\left(f_w\left(x^{(i)}\right)-y^{(i)}\right)x_j^{(i)} \tag{7.18}
$$

At first glance, the rationale behind (7.18) is not obvious. However, the logic underlying the equation becomes more apparent as shown in the following text box.

> new_weight = old_weight – learning_rate (actual_output – desired_output) * input

The new weight comes from taking the previous weight and changing it by a small amount, given by the difference between the actual output of the network and the desired output, scaled by the magnitude of the input. It should be obvious that if the actual output of the network matches the desired output, then the change to the old weight will be zero. If, however, the actual output does not match the expected output, then the change to the old weight will depend on (1) the magnitude of the difference between the actual output and the desired output, combined with (2) the magnitude of the input. An example is shown in Figure 7.4 to demonstrate the learning progress versus iterations.

7.1.3 Logistic Regression

We've seen how a simple linear model combined with the powerful tools of gradient descent and a suitable loss function can be applied to build machine

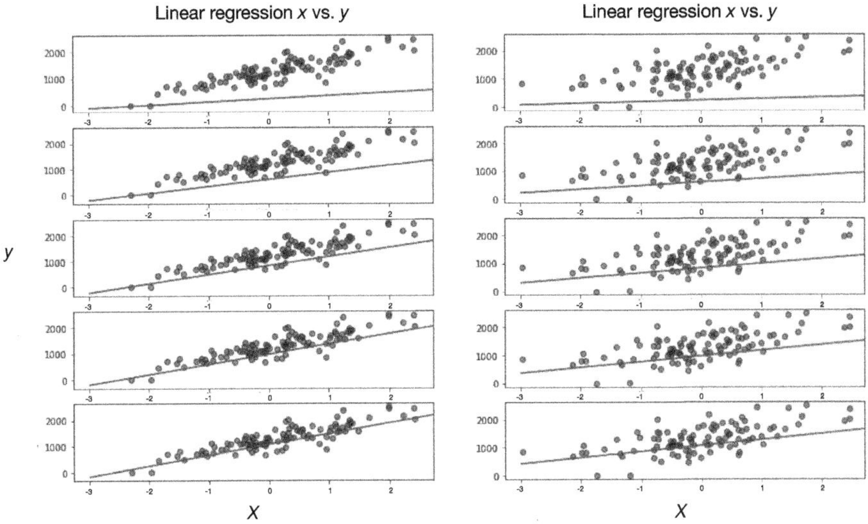

Figure 7.4 An example of applying gradient descent to learn the weights for the two-dimensional linear regression radar problem. The straight line shows the fitted line learned by the model. The top row shows the model learned after 10 iterations of gradient descent, with the final model learned after 100 iterations of gradient descent in the lowest row.

learning models that can fit and estimate linear regression. What if our objective was to estimate the discrete category or class of an input, such as to estimate if an input example contained a target of interest? While you could imagine simply reusing linear regression, fitting a line to the data, and then thresholding the output value at, say, 0.5 (with outputs >= 0.5 containing a target and outputs < 0.5 not containing a target), there is a better way.

We can achieve this with logistic regression, which despite its name, is a widely used tool for classification.

7.1.3.1 Logistic Regression Toy Problem

To motivate the discussion of logistic regression, let's assume that we received a software update from the ToyRadarProblems company for the same $1 radar sensor used previously for linear regression. The software update provides the option to operate the sensor in two modes: *distance mode* or *classification mode.* The sensor still generates $x \in \Re^2$: a two-dimensional signal, where x_1 represents secret radar feature one, and x_2 represents secret radar feature two. In distance mode, you observe that the x values look identical to those before the software update and can be used to estimate a user's distance in centimeters using the previously trained linear regression model. However, you discover that in classification mode the x values look very different and appear to correlate with the type of motion in front of the sensor. Based on this, you want to train a machine learning model to estimate if the object in front of the sensor is a person or a dog. To try this, you record a small dataset of 100 examples of sensor values from the radar paired with if you or your dog were in front of the sensor. We will denote recordings with a person as class 1 and recordings with a dog as class 0.

7.1.3.2 Logistic Regression Problem Definition

Logistic regression shares many attributes with linear regression. Like linear regression, we still want to model y as a linear function of x; however, as we are interested in binary classification, it is beneficial to represent the output of our model as the *probability* of x belonging to class 0 or class 1. Given the probability of an input belonging to each potential class, the discrete class label can be selected by thresholding the probability at a specific value (e.g., >= 0.5). As this is a binary problem, y is represented as: y = {0,1}. Similar to linear regression, we can simplify the notation of the linear model by adding the constant of 1 to the input vector at x_0 with the constant value of 1, giving $x = [1\ x_1\ x_2]^T$.

To map the real value generated by the linear combination of $w^T x$ to a probability in the range of {0,1} we can leverage a powerful function from statistics called the sigmoid function, also known as the logistic function:

$$g(z)=\frac{1}{1+e^{-z}} \tag{7.19}$$

where $z = w^T x$. Figure 7.5 illustrates the classic *S*-shaped curve of the logistic function.

The logistic regression model can therefore be defined as:

$$y = f_w(x) = g\left(w^T x\right) = \frac{1}{1+e^{-w^T x}} \tag{7.20}$$

7.1.3.3 Logistic Regression Cost Function

The cost function of linear regress cannot be applied to logistic regression in that it is nonconvex with many local minimums and can trap the algorithm from reaching the optimal solution.

Instead, the cost function of logistic regress is defined as:

$$J(w)=\frac{1}{M}\sum_{i=1}^{M}\left(y^{(i)} log(f_w\left(x^{(i)}\right)\right)+ (1-y^{(i)}) log(1-f_w\left(x^{(i)}\right))) \tag{7.21}$$

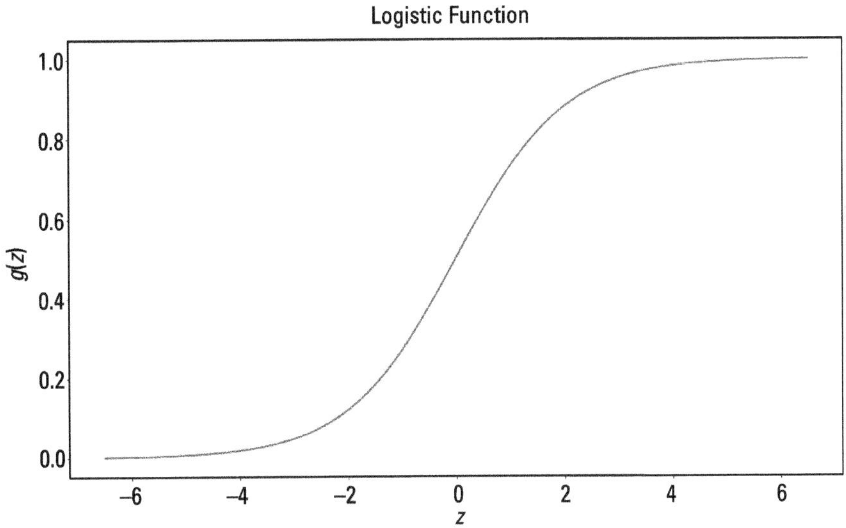

Figure 7.5 The logic function is an S-shaped curve.

What's surprising is that if we follow the same steps to compute the partial derivative of the logistic regression cost function as described for linear regression, we arrive at the same weight update rule: $w_j := w_j - \eta \sum_{i=1}^{M} \left(f_w \left(x^{(i)} \right) - y^{(i)} \right) x_j^{(i)}$. If we compare this to the weight update rule in (7.18) it looks identical; however, it should be noted that this is not the same algorithm, as $f_w(x^{(i)})$ is now a nonlinear operation for logistic regression.

7.1.4 Beyond Linear Models

The previous section described in detail several fundamental machine learning concepts—such as cost functions, gradient descent, and learning rates—through the lenses of linear and logistic regression. While understanding the fundamental theoretical concepts behind linear and logistic regression is important, in practice linear and logistic regression are rarely used for real-world sensing problems, like computer vision or radar.

Instead, more advanced machine learning techniques are required that can handle the complexities of real-world data. Understanding the relationship between basic linear models, such as linear regression, and more sophisticated nonlinear techniques, such as neural networks or support vector machines is a key part in successfully training and deploying robust machine learning models.

The key property that can be used to distinguish between classic machine learning techniques and more modern techniques is:

> Can the machine learning technique automatically learn new feature representations from the original data representation that map a difficult high-dimensional nonlinear problem to a simple linearly separable problem?

Classic machine learning techniques such as linear or logistic regression cannot. Indeed, even more advanced machine learning techniques such as support vector machines [10] or random forests [11], which support nonlinear machine learning problems and were considered state of the art throughout the 1990s and until around 2012, still cannot learn how to transform the rich, high-dimensional data typical with computer vision or radar sensors into simple linearly separable problems. These more classic machine learning techniques are strongly dependent on the types of feature engineering used in combination with the underlying machine learning algorithm. Feature engineering techniques are required to transform the original data into a representation that fundamentally improves the performance of the machine learning algorithm, as more classic machine learning algorithms lack the ability to automatically learn these representations directly from the data. Examples of hand-designed feature engineering techniques for computer vision include algorithms such as histogram of gradients [12], scale-invariant feature transform (SIFT) [13], and speeded up robust features (SURF) [14].

At this time, the leading machine learning algorithms for complex real-world sensing data are based on modern neural network techniques [3, 8]. This is because neural networks, particularly models with many layers—often called deep neural networks—can learn how to automatically transform complex high-dimensional nonlinear data that is input to the network into a simple linearly separable subspace at the output of the network. In other words, the beauty of modern neural network techniques is that they automatically learn the features required to transform a complex real-world signal—like a camera image or a radar signal—input to the network, into a new representation that can be separated with a linear model at the final layers of the network.

7.1.5 Neural Networks

Artificial neural networks represent a diverse family of techniques for machine learning that are widely used throughout science, medicine, and engineering. Indeed, one key advantage of neural networks, particularly deep neural networks, is that breakthroughs in one domain (e.g., image classification) can rapidly be applied to other domains (e.g., speech recognition, medicine, creative arts, fraud detection, material design, human computer interaction). Likewise, innovations in one sensing modality, such as computer vision, can rapidly scale to orthogonal sensing techniques, such as audio, inertial sensors, and most importantly radar.

7.1.5.1 Feedforward Networks

The simplest form of artificial neural network is the feedforward network [15], a specific category of neural network where connections between the nodes do not form a cycle.

Feedforward networks consist of an *input layer*, which propagates data through zero, one, or more *hidden layers*, followed by a final *output layer*. Each hidden layer and output layer consists of nodes that take the input from the previous layer, apply some form of processing, and output a single scalar that is propagated onto the nodes in the following layer. Figure 7.6 illustrates a basic feedforward network with one input layer, two hidden layers, and an output layer.

The elegance of neural networks is that very powerful functions can be constructed by chaining together a large number of incredibly simple operations. This is similar to how complex digital sounds, such as a trumpet, can be constructed by summing thousands of sine waves with varying frequencies and phases. Indeed, neural networks have been shown to theoretically represent any arbitrary function via multiple universal approximation theorems.

7.1.5.2 Perceptron Feedforward Networks

The simplest form of feedforward network is the perceptron algorithm, illustrated in Figure 7.7.

The perceptron is a binary classifier that dates back to 1958 [16]. The perceptron outputs the value 1 if the result of the dot product between the input and its weight vector is greater than zero; outputting zero otherwise:

$$f_w(x) = sign\left(w^T x\right) \tag{7.22}$$

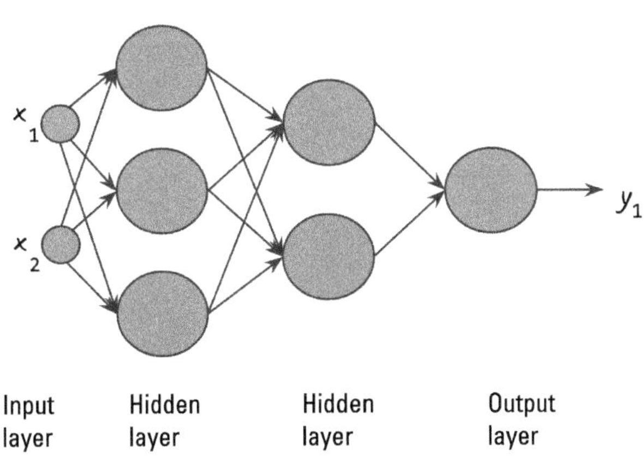

Figure 7.6 An illustration of a basic feedforward neural network with two inputs, two hidden layers with three units and two units respectively, and an output layer with a single output.

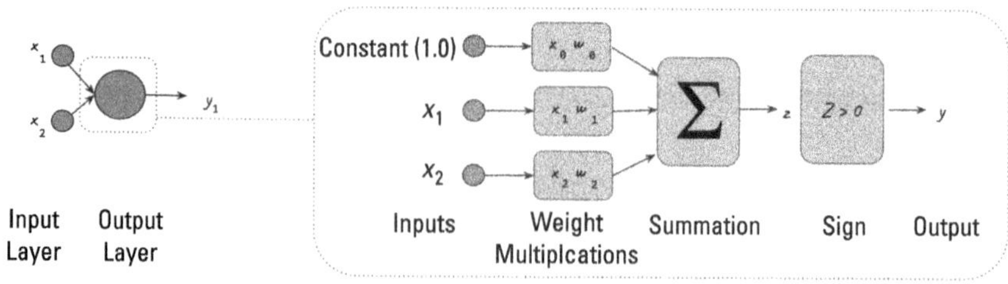

Figure 7.7 An illustration of a single-layer perceptron showing the internal logic applied to the two inputs to generate a single output.

where x is the N-dimensional input vector, with the convention used earlier in this chapter of the constant 1 at element zero; w is the N-dimensional weight vector, with the convention of weight-element zero representing the bias term; and sign(·) is the sign operation that outputs 1 if the input is greater than zero; zero otherwise.

Equation (7.22) shows that the perceptron is very similar to the linear and logistic regression algorithms described earlier in this chapter. Like linear and logistic regression, the perceptron has a single weight vector capable of learning one decision boundary, combined with a simple thresholding operation via the sign operator—with the result that any dot product between $w^T x$ that is greater than zero outputting a one, or a zero otherwise.

In 1969, Marvin Minsky and Seymour Papert demonstrated that a single perceptron is only capable of solving linearly separable problems [17]. The perceptron algorithm is therefore not very useful, as most real-world machine learning tasks are not easy linearly separable problems. Nonetheless, very powerful, nonlinear models can be constructed by chaining together multiple layers of these fundamental building blocks to form one of the key components of most modern machine learning algorithms: multilayer perceptrons.

7.1.5.3 Multilayer Perceptron Feedforward Networks

A multilayer perceptron (MLP) consists of a chain of fully connected units with one or more hidden layers, as shown in Figure 7.8.

Each layer in the MLP consists of one or more units—also known as neurons or nodes. Each unit takes an N-dimensional input, x, and outputs a scalar value,y. The outputs from layer l then become the inputs for all the units in layer $l+1$ which output new y values, and this is repeated until the final output layer is reached.

In an MLP, each unit across all layers performs the same general operation, given by:

$$f_l(x) = g\left(w^T x\right) \quad (7.23)$$

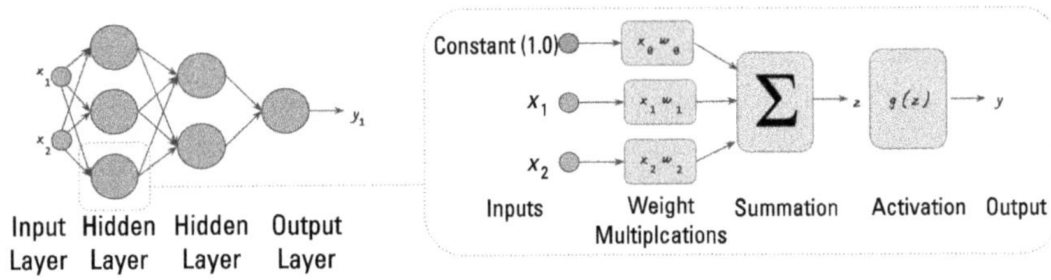

Figure 7.8 An illustration of a multilayer perceptron with one input layer, two hidden layers, and an output layer. The internal logic highlighted for one unit is the same across all units in the network.

where x is the N-dimensional input at layer l, w is the weight vector for the specific unit at layer l, and $g(\cdot)$ is an activation function, which can apply either a linear or nonlinear operation on the dot product between $w^T x$. Figure 7.9 illustrates four common activation functions.

As the output of one layer is fed as input to the next layer, an MLP can be thought of as a chain of n recursive operations:

$$f(x) = f_0\left(f_1\left(f_2 \ldots \left(f_n(\bullet)\right)\right)\right) \tag{7.24}$$

As there are one or more units within each layer of a MLP, the dot product between the layer's input, x, and a unit's weight vector, w, can be also viewed as a matrix operation:

$$z = Wx \tag{7.25}$$

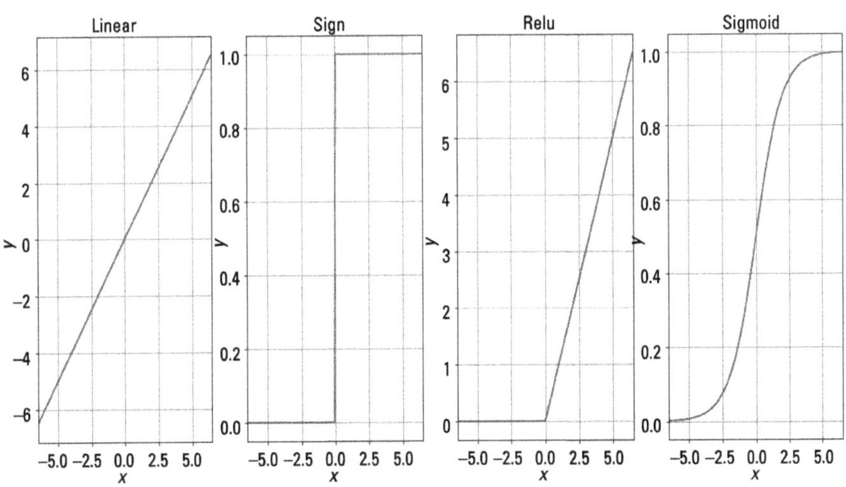

Figure 7.9 An example of four activation functions. From left to right these are linear, sign, rectified linear unit (ReLU), and Sigmoid.

where z is the output of layer l before any nonlinear processing via the activation function, x is the N-dimensional input vector to layer l, with the convention of a constant 1 inserted at the element zero, and W is the weight matrix for layer l with M rows and N columns. Each of the M rows corresponds to one of the M units in the layer. An example of the operation is shown in Figure 7.10.

7.1.5.4 Solving XOR Using Multilayer Perceptrons

The difference between the single-layer perceptron shown in (7.22) and a single MLP unit shown in (7.23) is subtle; however, the result is profound. By chaining together multiple layers, complex nonlinear problems can be solved.

A simple example of this expressive power is demonstrated via the {AND, OR, XOR} binary logic problems. Each binary logic problem can be viewed as a classification problem with four points located at the coordinates {0, 0}, {1, 0}, {0, 1}, and {1, 1}, as shown in Figure 7.11.

In the AND problem, both input values must be 1 for the class label of 1 to be assigned, resulting in only the top right {1, 1} point being assigned to class 1 with all other points assigned to class 0. In the OR problem, either input value can be 1 for the class label of 1 to be assigned, resulting in {1, 0}, {0, 1}, and {1, 1} points being assigned to class 1 and only {0, 0} assigned to class 0. However, in the XOR problem, points are assigned the class label of 1 if the input values do not match; otherwise, the class label of 0 is assigned. This results in {1, 0} and {0, 1} being assigned to class 1 with {0, 0} and {1, 1} being assigned to class 0. Unlike the AND and OR binary problems, which can be solved with linear solutions, XOR is a nonlinear classification problem and therefore cannot be solved with a linear model.

Figure 7.12 illustrates how the AND and OR logic problems can be solved with a linear classifier. Solving the XOR problem requires a nonlinear classifier. This can be solved by adding a hidden layer in the MLP architecture with two units in the hidden layer, as shown in Figure 7.12. Each unit in the hidden layer

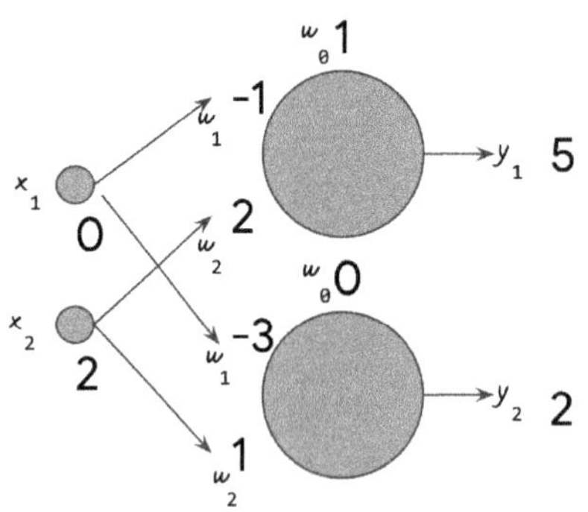

$$\begin{bmatrix} 1 & -1 & 2 \\ 0 & -3 & 1 \end{bmatrix} \begin{bmatrix} 1 \\ 0 \\ 2 \end{bmatrix} = \begin{bmatrix} 5 \\ 2 \end{bmatrix}$$

Figure 7.10 Example of MLP operations before activation function.

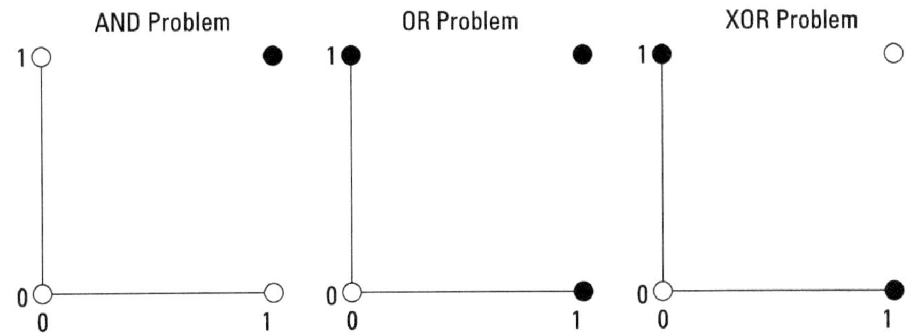

Figure 7.11 An illustration of the AND, OR, and XOR logic problems. AND and OR are linear problems and can be solved with linear classifiers. XOR is nonlinear and requires a nonlinear solution.

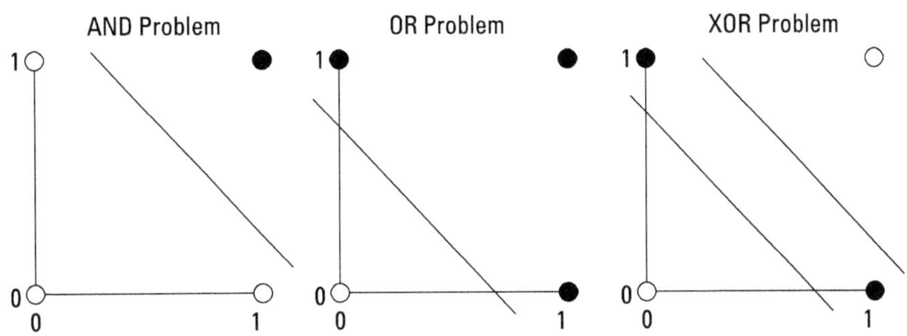

Figure 7.12 An illustration of potential solutions to the AND, OR, and XOR problems.

can solve a complementary subproblem, which can then be combined in the output layer to solve the main XOR problem.

For example, if the first unit in the hidden layer learns to solve the OR problem (which we know is linear and can be solved with one unit), and the second unit in the hidden layer learns to solve the NAND (not AND) problem (which we also know is linear and can be solved with one unit), then the resulting output of these two hidden units can be fed as input to the output layer, which just needs to solve the AND problem (which we know is linear and can be solved with one unit). In other words, we can take a harder nonlinear problem and break it down into simpler linear subproblems, and then combine the results to get a nonlinear solution. The process is illustrated in Figure 7.13.

7.1.5.5 From XOR to Real-World Problems

The previous section described how a simple MLP with one hidden layer and two hidden units can be used to solve the XOR binary classification problem. What is incredible about neural networks (and many machine learning

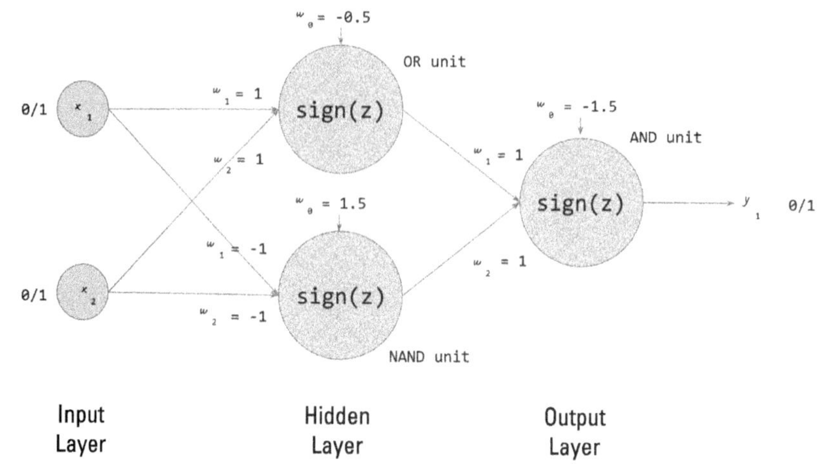

Figure 7.13 An illustration of one potential solution to the XOR problems using a multilayer perceptron with one hidden layer. The two units within the hidden layer each solve the OR and NAND logic problems, with the output layer combining the answers of the hidden layer in an AND operation to achieve the solution to the XOR problem.

algorithms) is that the key principles involved in solving toy problems such as XOR can be scaled up to provide solutions for extremely complex real-world challenges, such as image classification, natural language processing, and radar target detection.

Indeed, one of the reasons that neural networks can solve difficult classification and regression tasks is that they learn to transform complex and highly nonlinear problems in the input layer of the network into well separated linearly separable problems at the upper layers.

Figure 7.14 shows an example of how a neural network like an MLP can start to scale from solving basic linear classification problems that require a single decision boundary, to more demanding problems that require multiple decision boundaries to create decision regions. The right-most example in Figure 7.14 shows a more subtle nonlinear problem that requires the combination of multiple hidden units to define the higher-order decision boundary.

7.1.5.6 Activation Functions

A key property of neural networks is their capacity to represent complex, nonlinear functions. Indeed, one of the reasons that neural networks can solve difficult classification and regression tasks is they chain together multiple nonlinear layers that transform complex, highly nonlinear problems in the input layer and lower layers of the network into well separated linearly separable problems at the upper layers. This is achieved through activation functions. Common activation functions are listed next.

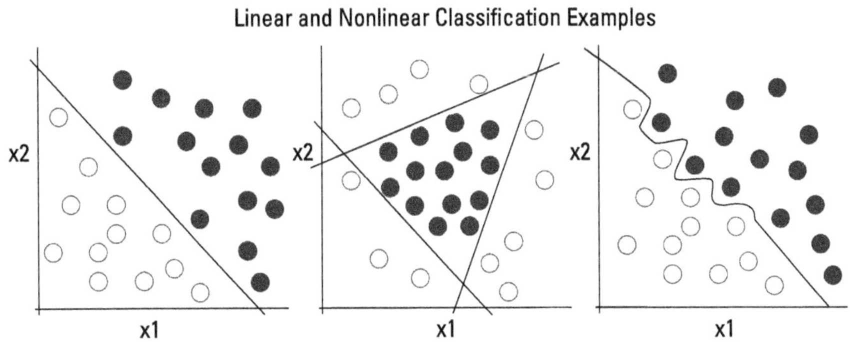

Figure 7.14 An illustration of a linear and two nonlinear classification problems. From left to right: (i) a basic linear classification problem that can be solved with a single decision boundary; (ii) a nonlinear classification problem that can be solved with a MLP with one hidden layer with three hidden units, with each unit learning one of the decision boundaries; (iii) and a nonlinear classification problem that requires a larger number of hidden units to represent the complexities of the decision boundary.

Linear

The linear activation function is the simplest form of activation function, as it simply returns the input.

$$f(z) = z \tag{7.26}$$

Rectified Linear Unit

ReLU is a piecewise linear function that outputs the input value directly with no modifications if it is positive; zero otherwise.

$$g(z) = max(0, z) \tag{7.27}$$

In many applications, ReLU has become the default activation function for MLP and convolutional neural networks due to its simplicity and speed. ReLU is particularly useful in neural networks with many layers, as it is less sensitive to the vanishing gradient problem that is prominent in other nonlinear activation functions.

Sigmoid

The Sigmoid function, also known as the logistic function, is the same function described earlier in the context of logistic regression:

$$g(z) = \frac{1}{1+e^{-z}} \tag{7.28}$$

While in practice the Sigmoid function has been replaced by the ReLU function, it is still an important activation function to be aware of.

Softmax

The Softmax function is a generalization of the Sigmoid or logistic function to multiple dimensions. The Softmax function is a common activation function to use in the output layer of any neural network architectures applied to classification. This is due to the Softmax functions taking an input vector of K real numbers and normalizing this into a probability distribution consisting of K probabilities proportional to the exponentials of the input values.

$$g(z)=\frac{e^{-z_i}}{\sum_j e^{-z_j}} \tag{7.29}$$

7.2 Radar Machine Learning

In this section, we discuss how machine learning can be successfully applied to radar for target detection, gesture sensing, and beyond.

7.2.1 Machine Learning Considerations for Radar

While modern machine learning techniques have the advantage of wide adoption across sensing domains, there are specific properties of radar that require careful attention.

For example, in computer vision, certain assumptions can be made regarding the adjacency of neighboring pixels being semantically related [2]. This assumption is exploited by the convolutional layers within a convolutional neural network [18] with unprecedented success in computer vision. However, the structure of radar data can frequently break these adjacency assumptions depending on the radar signal representation being used—with the dimensions in radar transformations more commonly representing temporal, range, or velocity components of the underlying signal. Likewise, when a target moves two times in range further from a radar sensor, the target's signal power will drop 16 times, according to radar range equation.

We list ten properties of radar that require attention when designing radar machine-learning systems. Guidance on each of these considerations is provided throughout this chapter.

Consideration 1: Radar Hardware

There are many categories of radar (e.g., pulse, CW, FMCW, FSK), but not all radar systems provide the same information or resolution. While there are techniques that provide hardware abstraction layers to generalize radar hardware [5], care should be taken to ensure the hardware selected for a specific machine learning task provides the necessary information and resolution for that task (e.g., radar has the necessary range resolution, angular resolution, and provides the SNR required for a given task). Likewise, bringing up an algorithm or machine learning model on one category of radar does not necessarily guarantee that the algorithm or model can be applied to other radar hardware categories—even with significant tuning or retraining for some hardware categories due to fundamental differences in signal or resolution.

Consideration 2: Radars Are Not Cameras

Radar, particularly broad antenna beam strategies, such as those used in wearable sensors [5], has more in common with audio signals than images. While it's certainly possible to directly apply state-of-the-art techniques originally optimized for camera-based images or time-of-flight sensors to radar, care should be taken when designing radar-based machine learning algorithms. For example, radar has excellent range resolution (~2 cm for 60-GHz radar based on current FCC regulations); however, broad antenna beam strategies can be limited to low cross-range spatial resolution. Therefore, trying to apply state-of-the-art techniques designed for sensors with higher spatial resolution may fail for specific use cases.

Consideration 3: Temporal Processing (Slow Time)

One of the key strengths of radar is its capacity to accurately detect targets moving at incredibly high speeds. While it is possible to infer information from a single radar time instance, the real strength in radar processing comes from temporal processing over subsequent time frames. Temporal processing can be used to boost the signal-to-noise ratio of weak targets or to infer temporal characteristics of a target to aid in classification, such as in gesture detection. The length of the time sequence will be context dependent. Nonetheless, applying signal processing and machine learning techniques that can leverage this temporal domain can significantly boost algorithm performance and unlock key information in the radar signal, such as Doppler data.

Consideration 4: Clutter

Clutter is the unwanted background scatter generated by other objects within range of the sensor that adds complexity to the main signal of interest from the target or targets the radar is attempting to detect. Removing clutter from the radar signal is similar to the task of background subtraction in computer vision.

Clutter is typically removed from the radar signal via filtering techniques in the early stages of the signal processing chain, as discussed in previous chapters. However, depending on the use case, it may require a machine learning algorithm to learn how to remove clutter for more complex use cases.

Consideration 5: Complex Signals

Many radar sensors provide signal representations that are best analyzed as complex signals (i.e., real and imaginary pairs). Complex signal analysis is common in the signal processing community; however, it is less frequently used in the field of machine learning as the most common data types are all real-valued. While a complex signal can naively be treated as a pair of floating pointing values and input to a standard machine learning model with some success, there is an opportunity to advance the field of complex-based machine learning models for radar, and in turn potentially improve the quality, diversity, and impact of radar algorithms.

Consideration 6: Radar Data Transformations

There is a substantial body of research in the signal processing community for advanced data transformations for radar. Transformations such as range Doppler, range profile; micro Doppler, and fast time spectrograms are a few examples of custom data transformations that can be applied to the raw radar signal output by radar hardware to help boost and filter the original signal.

The machine learning community can leverage these transformations to help filter out unwanted signals, such as background clutter, and boost the signal-to-noise ratio of key targets in the signal. Machine learning researchers should familiarize themselves with the common transformations that can be used to filter and boost raw radar signals. They should also pay attention to transformations that add little value for a given use case, as radar signal processing transformations can be high dimensional and are typically sparse.

Consideration 7: Data Augmentation

Data augmentation is a fundamental part of any modern machine learning workflow. Machine learning researchers should pay particular attention to the type of augmentation strategies applied to radar data, as classic data augmentation techniques used in the computer vision community may not be suitable for radar data. For example, common augmentation techniques used in the computer vision community include randomly flipping an image, applying random rotations to an image, applying random shearing to an image, applying random brightness or contrast changes, and so forth. These techniques are not suitable for radar data and may in fact degrade the performance of a radar-based machine learning, rather than enhance it.

Consideration 8: Limited Radar Datasets

A significant number of machine learning breakthroughs over the past decade have been directly linked to large, high-quality, publicly available benchmark datasets. Examples include imagenet [9], MNIST, and fashion-MNIST. Benchmarks such as imagenet enable machine learning researchers to fairly A/B compare two or more algorithms against a common dataset and minimize the substantial overhead, costs, and time associated with manually creating and annotating a custom dataset.

Unfortunately, similar large, high-quality, publicly available benchmark datasets are lacking for radar at this time. This makes it difficult for researchers to advance the state of the art in radar techniques and to benchmark new algorithms on common reference datasets.

Consideration 9: Manual Labeling Challenges

All machine learning techniques have a strong dependency on high-quality ground truth labels. Even if labels are not required for training models, high-quality ground truth labels are still necessary to accurately test and benchmark a model's performance.

Building large, high-quality, labeled datasets is a nontrivial, time consuming, and expensive task. Moreover, providing ground truth labels to a dataset is hard enough in domains like images or audio where a nonexpert can manually label the majority of a dataset. However, labeling radar datasets adds additional challenges as it can be difficult for even expert reviewers to view a radar recording and accurately annotate the data with ground truth labels.

Consideration 10: Radar Data Formats and Radar Libraries

The field of radar is missing the mature data formats and data processing libraries that are commonplace in other domains, such as computer vision or audio processing. While there have been recent efforts to standardize radar data formats and open source radar libraries, such as OpenRadar [19], more progress in this area will significantly help to accelerate radar machine learning.

7.2.2 Gesture Classification

Gesture recognition has a distinct set of requirements that make it particularly applicable for radar-based recognition algorithms.

Gesture recognition sensors should ideally:

- Be small enough to fit within wearable devices, without requiring an aperture or cutout in the devices exterior;
- Be invariant to lighting, temperature, and other environmental conditions;

- Operate at low power to support multiday-use battery life;
- Support always-on use cases while maximizing user privacy;
- Support touchless interactions at either near field or far field;
- Support the detection gestures even when parts of the hand or body are partially occluded from the sensor.

Furthermore, gesture recognition algorithms should:

- Work robustly across a diverse range of users of all ages, body sizes, mobilities, genders, and ethnicities;
- Detect gestures with minimal latency to support frictionless user experiences;
- Fit within the aggressive compute and memory constraints of low-power embedded systems;
- Support high detection rates while not generating false triggers due to random motions near the sensor.

Satisfying all these requirements is nontrivial. Indeed, many of these requirements are directly opposing. For instance, to minimize latency in gesture predictions, detection events need to be generated as a gesture is being performed; however, this significantly increases false triggers, as many random motions look very similar to valid gestures and only at the end of the motion is enough context available to robustly differentiate a true gesture from harder background motions.

There are additional challenges that come from touchless gesture sensors, the most significant being that gestures need to be detected from an unsegmented continuous stream of sensor data. Unlike with a touch screen, where a user physically touches the screen to create a clearly distinguishable touch event, touchless gesture sensors like radar don't explicitly know when a gesture starts or ends. Instead, they must continually search for gestures and detect a true gesture event from thousands of random nongesture motions that occur within range of the sensor. This problem is known as *gesture spotting*.

An additional challenge in gesture recognition is the detection of dynamic gestures vs the detection of static postures. Generally speaking, detecting static postures is an easier task, as the posture can be detected from a single time instance or predictions from coherent time instances can be averaged to increase detection confidence. Alternatively, dynamic gestures have a strong temporal component and may be performed at significantly different speeds by different users. Indeed, detecting dynamic gestures in low-power wearables can be infeasible for many sensors, as the motion exceeds the sensors' Nyquist frequency and lacks information to accurately recognize dynamic gestures. However, detecting fast-moving objects is one of the strengths of radar, which makes it particularly useful for dynamic gesture recognition—even under the constraints of low-power wearable devices.

A further challenge in gesture recognition is the large variability in how different users perform the "same" gesture, with this variability typically increasing as the number of physical constraints of the system decrease. This natural variability can occur from the user's physical attributes (e.g., the size of their hand in a gesture detection system or their height in a presence detection system); cultural or social attributes (e.g., gestures may need to be changed in some countries or contexts, such as the "OK" sign in Western culture, which could be found offensive in Brazil); medical or health reasons; or a myriad of additional reasons.

The most significant challenge in any touchless gesture recognition system is balancing the tradeoff between robustly detecting as many true gestures as possible, while simultaneously not false triggering on random motions performed within range of the sensor. While the tradeoff between a high gesture detection rate and achieving a low false positive rate can be adjusted for each use case; attaining good gesture recall while minimizing falsing across a diverse range of users and scenarios requires significant attention to detail.

To summarize, the robust detection of dynamic gestures requires solving three nontrivial problems:

1. Gesture spotting: how to detect a gesture from a continuous stream of nonsegmented sensor data;
2. Temporal sequences: how to detect a nonstatic motion that evolves over a variable time sequence;
3. Balancing gesture detection versus false triggers and ensuring this works across a diverse range of users.

Modern gesture recognition techniques typically address these three problems holistically with a single gesture detection algorithm. Solving these problems holistically allows a single machine learning model to learn the important relationships between, for instance, detecting noisier gestures and rejecting harder background motions that look similar to true gestures.

7.2.2.1 Gesture Recognition Pipeline

Gesture recognition techniques commonly consist of the following three stages (Figure 7.15) of data processing chained together to form a gesture recognition pipeline [5, 20]. The input to the pipeline consists of the radar data output di-

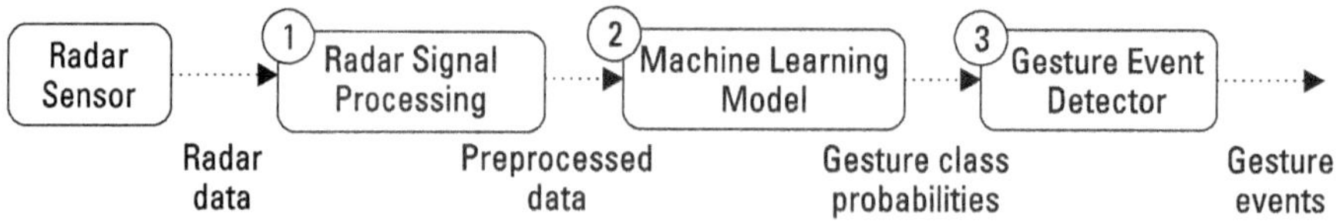

Figure 7.15 An illustration of the three main stages of a common gesture recognition pipeline.

rectly by the hardware (or read from a recorded file), with discrete gesture events being output from the final stage of the pipeline.

Radar Data

The radar sensor outputs encoded radar samples via a digital software interface. Radar data can either be output as a continuous stream (e.g., 2,000 frames per second) or grouped into bursts of data (e.g., 30 bursts a second, with each burst containing *N* frames generated at 2,000 frames per second). Figure 7.16 illustrates two sampling strategies for frequency modulated continuous wave radars based on either (1) continuous sampling or (2) burst sampling.

Stage 1: Radar Signal Processing

The first stage in a radar-based gesture recognition pipeline is radar signal processing. The signal processing techniques applied at this stage are typically customized for each radar hardware and use case; however, the goals of this stage are the same: to boost the signal-to-noise ratio of motions of interest and to filter and remove unwanted signals, such as stationary targets.

The radar signal processing stage of a gesture recognition pipeline takes into consideration four of the ten radar properties highlighted in the previous section, with a specific focus on radar hardware, clutter removal, complex signal analysis, and radar data transformations. A great example of these

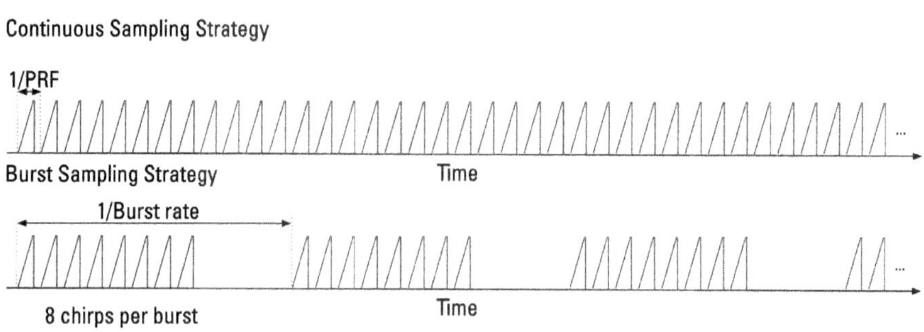

Figure 7.16 An illustration of a continuous sampling strategy and a burst sampling strategy for frequency modulated continuous wave radar.

components being applied in a holistic manner is the complex range Doppler (CRD) transformation.

CRD is a ubiquitous radar signal transformation technique that projects a temporal sequence of individual radar chirps into a new representation that represents the received radar signal as a function of range and velocity, as explained in previous chapters. An example CRD of a moving hand is shown in Figure 7.17.

The CRD transformation has multiple properties that bring value for gesture recognition and other use cases:

- *Range and velocity:* the CRD transformation provides a simple mapping to the range and velocity of one or multiple targets;
- *Multitargets:* the CRD transformation supports the detection of multiple targets (e.g., multiple fingers, multiple hands, and multiple body parts);
- *Signal boost:* the CRD transformation provides signal boost through the coherent integration of the complex radar signal across a coherent sequence of radar samples;
- *Clutter removal:* the CRD transformation can work with another processing stage that removes basic clutter signals from stationary environment via a high pass filter.

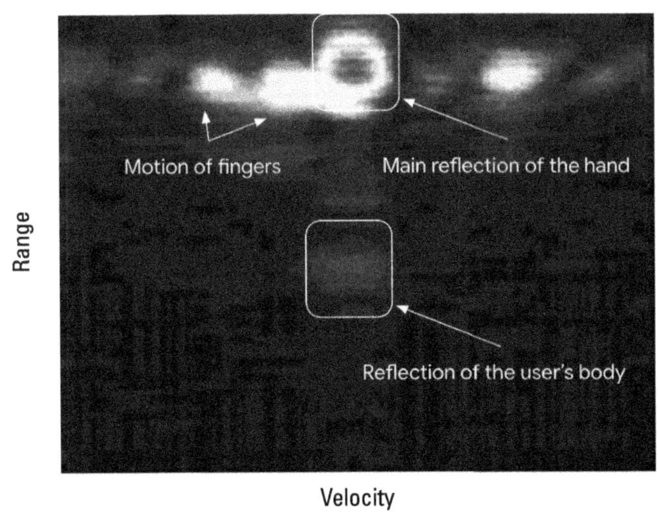

Figure 7.17 An illustration of the magnitude of a single channel of complex range Doppler data showing the motion of a hand while a user performs a swipe hand gesture.

The range and velocity resolution of each cell in the CRD transformation are given by the radar system parameters, as discussed in previous chapters.

To leverage the CRD transformation for gesture recognition, a stream of CRD data can be computed from a sequence of radar data, with the resulting CRD data fed as input to the machine learning model in the second stage in the gesture recognition pipeline. An example sequence is shown in Figure 7.18 for the CRD of a moving hand.

Stage 2: Machine Learning Model

The second stage in a radar-based gesture recognition pipeline is a machine learning model that converts a time series of radar signal processing transformations into gesture class probabilities.

Segmented and Unsegmented Gesture Recognition

While it is possible to detect basic targets from a single radar sample at a specific time instance, more complex tasks such as gesture recognition require a machine learning model to be fed with a sequence of radar samples that provide enough temporal context of the gesture that is performed.

As touchless gesture sensors don't explicitly know when a gesture starts and ends, this requires recognition algorithms to enable detection of gestures directly from a continuous stream of data. Artificially designed starts and ends are rarely used due to the related user friction. It is worth noting that supporting unsegmented gesture detection can make the task of gesture recognition significantly more challenging, as the model has to detect the gesture from a continuous input stream, which can contain thousands of other motions made within range of the sensor that are similar to the gesture.

One strategy to obtain unsegmented gesture recognition is to continually buffer a short time window of data and have a machine learning model classify gestures from the contents of the buffered time data [20]. Figure 7.19 illustrates this buffering concept, showing an example of buffering three radar time slices into a fixed sized buffer and updating the buffer every time step. In practice, the size of the buffer will be use case dependent, ranging from around one second for gesture detection examples to much longer time windows for other use cases such as activity recognition.

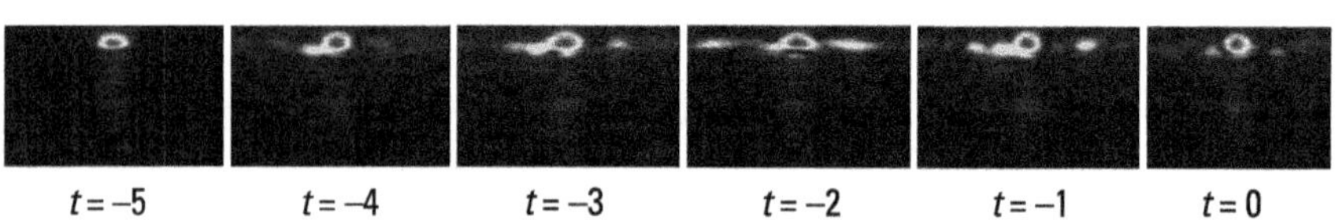

Figure 7.18 An illustration of a sequence of complex range Doppler data showing the motion of a hand while a user performs a basic hand gesture. Each slice shows the magnitude component of a single channel of complex range Doppler data.

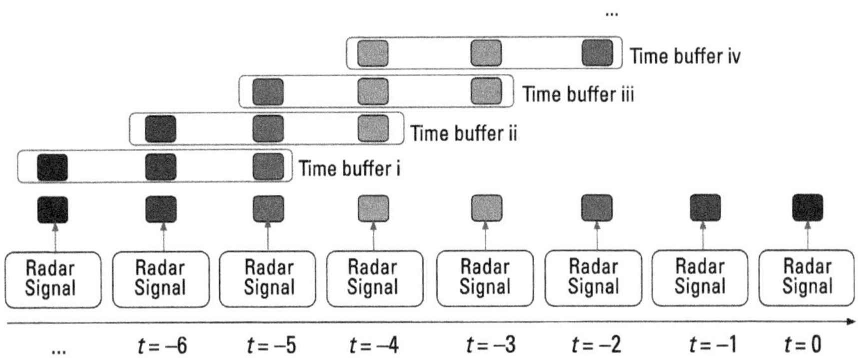

Figure 7.19 An illustration of buffering a time series of radar samples into a fixed length buffer.

Supporting a Nongesture Class

One important consideration in unsegmented gesture recognition is that a user will not always be performing one of the *K* gestures known to the model. It is therefore important to add an additional gesture class to the model that represents a *background class* [20]. The background class, typically denoted as class label 0, allows the model to output a prediction that is clearly not any of the known gestures if it receives an input sequence that does not contain any known gestures. It is common practice to refer to the class labels that represent gestures as *positive classes* and the generic background or nongesture class as the *negative class.*

Gesture Model

Given a time series of radar signal processing transformations, a machine learning model can be used to map this sequence into class probabilities representing the various gestures the model should recognize.

There are multiple machine learning strategies for handling time series sequences data, ranging from ignoring the temporal nature of the data and applying nontemporal techniques to dedicated time series algorithms such as dynamic time warping [21] or hidden Markov models [22].

Like most machine learning techniques, there has been a recent trend in shifting to neural-based techniques that can handle temporal data such as recurrent neural networks (RNN) like long short term memory (LSTM), with specific radar-based models such as DeepSoli [6] and RadarNet [7].

The advantage of neural techniques like LSTMs is that they are used as part of a larger neural architecture that can be trained holistically to extract low-level features from the input data, alongside mid-level temporal features via the LSTM, combined with high-level features required for the final gesture classification. An example of these different low-level, temporal, and high-level features is illustrated in Figure 7.20.

Figure 7.21 illustrates the low-level, temporal, and high-level layers for a radar-based neural network for gesture recognition, similar to the designs used in the DeepSoli and RadarNet models.

In both cases, the input to the neural network consists of a sequence of signal processing transformations such as range Doppler. The signal processing transformations are input to a *frame encoding*. The frame encoding extracts low-level feature representations from the input signal, typically using a series of stacked convolutional layers that pyramid the input data into a lower-dimensional vector representation.

It is important to highlight that the same frame encoding can be applied to each input in the input sequence, regardless of the input's temporal order within the sequence. Applying the same frame encoding across all inputs in a sequence enables the frame encoding to scale to arbitrary sequence lengths while retaining a fixed number of parameters or weights. Decoupling the number of trainable parameters from the length of the sequence helps to reduce the amount of data required to train the model while also acting as a form of regularization on model's weights, as longer sequences will have less opportunity to overfit if a different frame encoding was used on each unique input. An additional advantage of reusing the same frame encoding over a sequence of inputs is the number of weights that need to be stored in memory is significantly reduced—a critical requirement for low-power wearable devices that typically have resource-constrained memory and compute.

The output of each frame encoding is fed as a sequence as input to the *temporal encoding*. The temporal encoding takes into account the temporal relationship between the frame encoding output at time instance t, t_{-1}, t_{-2} and so on. Temporal layers such as LSTMs can be used to provide temporal encodings, such as the LSTM layers used in both DeepSoli and RadarNet models. Many temporal encodings will actually transform a sequence of inputs into a sequence of outputs, with the output of the temporal encoding at t_{n-1} being fed as an additional input to the temporal encoding at t_n along with the input sequence at t_n, as shown in Figure 7.22.

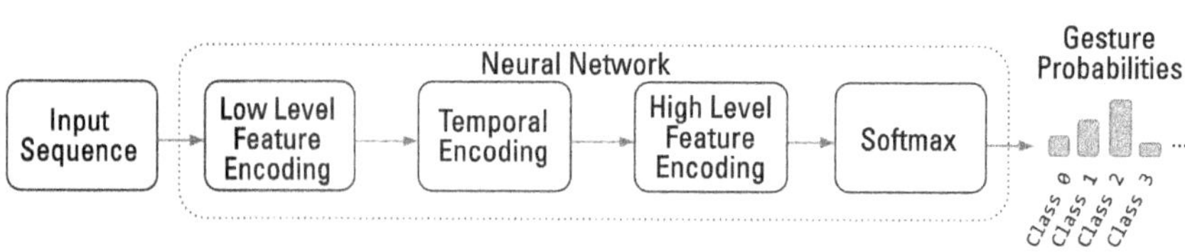

Figure 7.20 An illustration of the key building blocks of a neural network for gesture recognition.

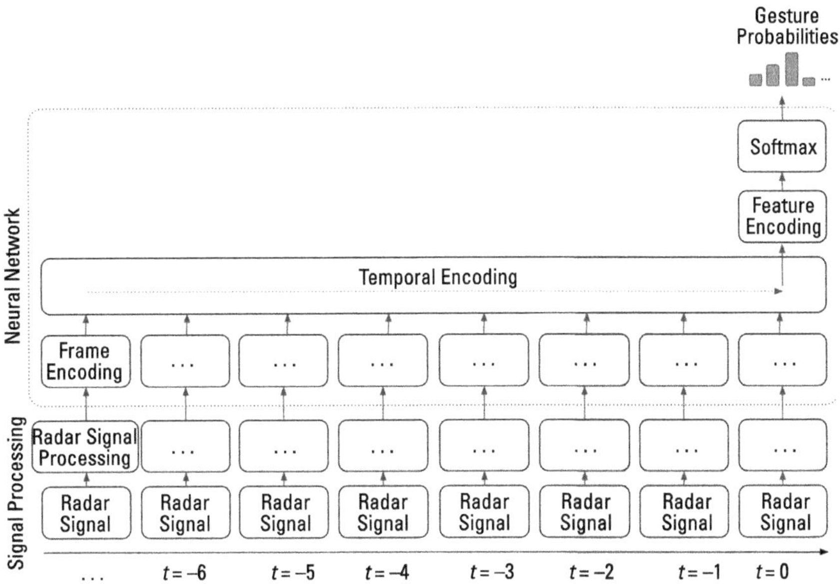

Figure 7.21 An illustration of the key building blocks for a radar-based neural network for gesture recognition.

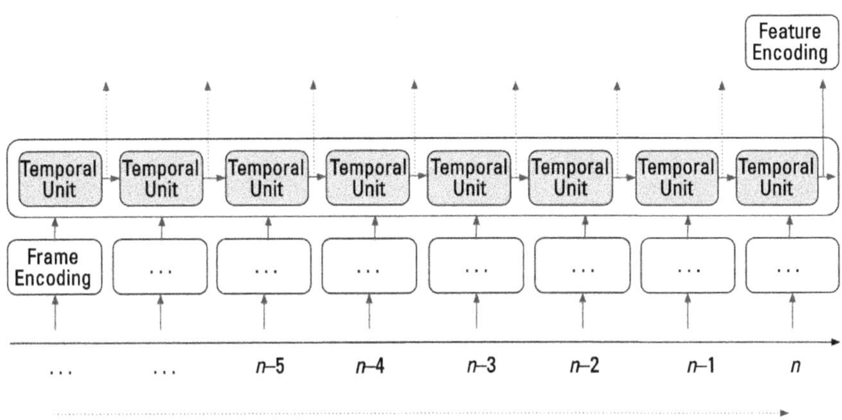

Figure 7.22 Process of temporal encoding.

The full output sequence of the temporal encoding can be used as input to the next layer in the neural network, although in practice only the final output from the temporal encoding at sequence step n is used as input to the next layer to help reduce the dimensionality of the model. Following the temporal encoding, there is typically one or more layers of feature encodings that extract high-level task-specific features from the output of the temporal encoder, with the final layer of the model being a Softmax layer. The Softmax layer normalizes the output of the neural network to provide a probability distribution over the predicted gesture classes.

Stage 3: Gesture Event Detection

The third stage in a radar-based gesture recognition pipeline is the gesture event detector. The detector converts the continuous gesture class probabilities output by the machine learning classifier in stage two of the pipeline into discrete gesture events.

Gesture events can be used as triggers by consuming applications or systems, such as triggering the next song to play in a music player if a "next" gesture is detected or canceling an alarm if a "skip" gesture is detected.

Discrete Gesture Events and Gesture Debouncing

The machine learning model in stage two of the radar-based gesture recognition pipeline outputs a set of normalized class probabilities, representing the model's confidence in each of the *K* respective gestures at that current time step.

The role of the gesture event detector is to transform the continuous probability output by the machine learning model into a discrete event that can be consumed by downstream applications. Figure 7.23 illustrates the continuous probabilities output by a machine learning model when a user performs an exemplar gesture along with the discrete gesture event that would be output by the gesture event detector.

The most trivial solution to transform the continuous gesture probability into a discrete event is via thresholding. A gesture event can be triggered anytime a positive class probability exceeds the detection threshold.

While naive thresholding is an option, it has multiple disadvantages, the most serious being *multiprediction errors.* A multiprediction error is when more than one event is triggered for the same gesture motion, even if each event has the correct gesture class prediction. An example is shown in Figure 7.24, where the same gesture triggered the event twice.

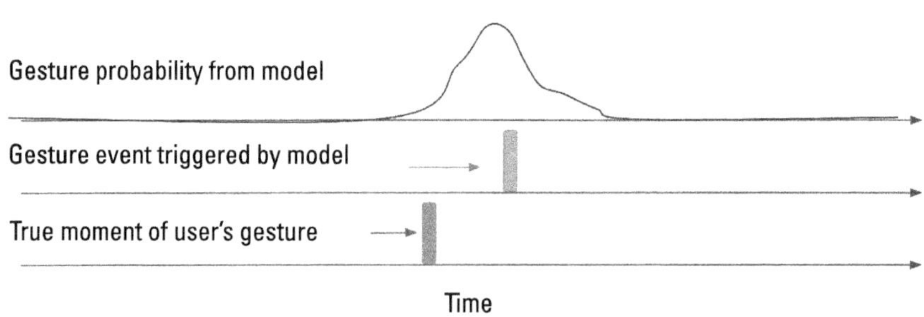

Figure 7.23 An example of the continuous probability output from a machine learning model and the discrete gesture event output by the gesture event detector.

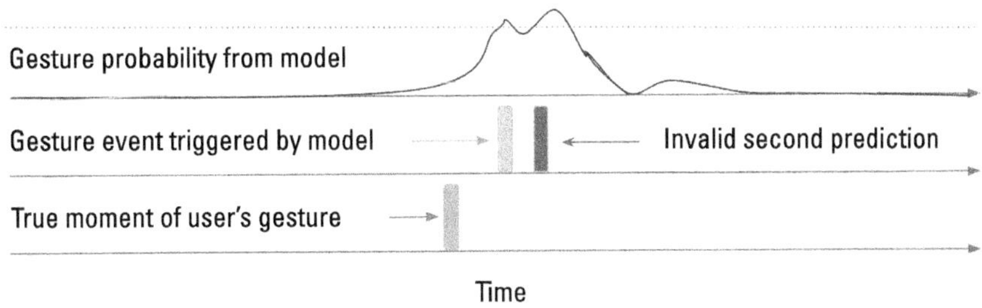

Figure 7.24 A second prediction is mistakenly claimed by the simple thresholding method.

For example, in a swipe classification task, a user performed one swipe right motion but the output from the gesture recognition pipeline emitted [SWIPE_RIGHT, SWIPE_RIGHT], instead of a single [SWIPE_RIGHT] event. Even if the multiprediction events are technically the correct class, they provide a frustrating experience for users, as it results in an action being triggered twice when the user intended the action to be triggered once.

To mitigate multiprediction errors, hysteresis can be added to the gesture event detector to mitigate multiple detections from a single motion. Hysteresis is a technique that intentionally applies a small delay to a system to avoid rapid changes in the system state, such as flickering in a user interface, or multiple gestures being triggered if the gesture probability fluctuates around the trigger threshold.

Hysteresis can be achieved by adding a second threshold to the gesture event logic, with the initial gesture event being triggered the instant the gesture probability exceeds the detection threshold, but with no additional events being triggered until the gesture probability drops below a second "reset" threshold.

Continuous Gesture Mapping

In the previous section, we described how a neural network can learn a mapping from a time series of radar data to discrete gesture events. To detect the discrete events, the neural network outputs a probability distribution over the K possible gesture classes, triggering a discrete event when the probability for a positive gesture class exceeds a detection threshold.

While the discrete detection of events is a ubiquitous application of machine learning to radar signals, machine learning models can do a lot more, including estimating continuous control parameters directly from radar data.

Continuous outputs could include the speed of a gesture, the range or position of the hand, or a myriad of other values. Moreover, the beauty of machine learning models, particularly neural networks, is that they can be designed to output both discrete and continuous predictions as needed for given use cases.

To enable a neural network to output continuous values instead of probability distributions, three small modifications are required:

1. *Continuous output head:* the default Softmax output used for classification should either be replaced or an additional output head should be added, with one output element for each continuous output value;
2. *Continuous loss function:* the loss function used in training should be modified to account for the continuous output signals;
3. *Continuous ground truth labels:* the ground truth labels should be adjusted to provide the necessary continuous targets for each output value.

Figure 7.25 shows an illustration of the main components of a neural network that can support both discrete gesture classification and continuous outputs. Note that the main backbone of the network is identical to that shown in the previous section for discrete gesture detection, with only an additional head added for the continuous regression.

To train a model that supports continuous outputs requires a nonclassification loss function. A common loss used for continuous targets is the mean squared error loss:

$$L(x,y,\theta)=\frac{1}{M}\sum_{i=1}^{M}(f_\theta\left(x^{(i)}\right)-y^{(i)})^2 \tag{7.30}$$

where $f_\theta(x^{(i)})$ is the estimated continuous output from the model given the ith input, and y is the expected target for the ith example.

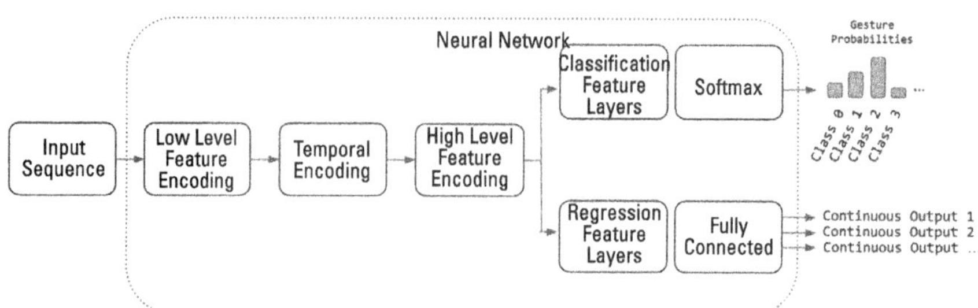

Figure 7.25 An example of a neural network with two output heads—one head for gesture classification and the second for continuous outputs.

7.3 Training, Development, and Testing Datasets

A central goal in machine learning is to learn models that not only perform well on the initial set of data used to estimate the model's parameters, but that can *generalize* to new examples. Generalizing to new unseen examples not in the initial dataset is nontrivial for real-world machine learning tasks.

To ensure a specific machine learning model is generalizing well, machine learning datasets are typically divided into three distinct groups: *training*, *dev* (also called validation), and *testing*. The goals of each dataset are as follows:

- *Training:* used to learn a model's parameters during the training phase with the goal that the model will generalize to new data not in the training dataset. A common practice is to train many machine learning models and select the "best" model based on some predefined metrics.
- *Dev:* used during the training phase to help monitor the performance of a model on data that is not in the training dataset. This can be used to help tune the hyperparameters[2] used to train the model or to down select a winning model from multiple candidate models being trained.
- *Test:* a specific dataset not shown to a model during the training phase or to help tune candidate models. Evaluating a specific model on the test dataset provides a more realistic estimate of the model's performance against unseen data.

7.4 Evaluation Methodology

There are many strategies to evaluate the performance of a machine learning model. Care should be taken to select the most suitable evaluation metrics and processes for each use case.

The following questions are helpful to frame and select what evaluation methodologies should be used:

- Is your use case classification, regression, or something else?
  - There is a wide range of metrics.
- Is the dataset balanced?

2. Hyperparameters are parameters whose value is used to control the learning process. Examples of hyperparameters are a model's learning rate, the number of filters in a specific layer in the model, the number of layers in a model, and the activation function used at a specific layer.

- Most real-world datasets are not balanced. For example, in a wearable product, a user might wear the product for 12 hours, but only interact with the product for a few minutes every hour. In this scenario, it is common to have a very imbalanced dataset with a small number of positive examples against a large number of negative examples.

- Are all errors equal to a user?
  - Users typically don't weigh false positive errors versus false negative errors versus misclassification errors versus other errors equally. Depending on the product or use case, the cost of a false positive error or false negative error could change significantly. Understanding how users view errors in your product and weighting them appropriately is critical to measuring and improving the quality of any product.
- Do your offline metrics reflect real-time performance?
  - Many factors can impact the data between offline metrics on a prerecorded test dataset and online metrics measured from using the same models in real time on device. These factors include differences in compute performance in offline and real-time systems, test coverage in offline datasets, full-stack system latency in real-time devices, the importance of real-time feedback, and many more. While there are many steps that can be taken to minimize the difference between offline and real-time metrics, the most important step is to not assume they are the same and measure both at regular intervals to understand how one metric maps to the other.
- How much do you trust your ground truth?
  - High-quality ground truth signals are critical for training and evaluating machine learning models. Even if nonsupervised techniques are used to train machine learning models, high-quality ground truth signals are still required to evaluate the performance of the models during testing. No ground truth system is perfect. Engineers should therefore understand the error inherent in whatever ground truth system they use and ensure that error is acceptable for the system they are building. For example, if the goal is to build a target tracking system capable of estimating the position of a user within 1 cm, but the error in the ground truth system is 10 cm, then clearly the accuracy of the ground truth cannot be trusted for evaluating the target tracking system's accuracy.
- Do you anticipate serving skew?
  - Serving skew is the phenomenon of fundamental differences in the performance of machine learning algorithms during training and de-

ployment. This discrepancy can relate to differences in how the data is handled or processed in the training and serving pipelines (e.g., per-batch normalization in training versus per-instance normalization in serving). Another common cause of serving skew is changes in the underlying data between training and serving (e.g., training data collected during the summer but serving in the winter). There are many potential reasons for serving skew, but the most important realization any machine learning engineer should recognize is that there will be some form of training/serving skew and they should (1) actively search for and monitor skew, and (2) put in place measures to mitigate and reduce skew where possible. Techniques to reduce serving skew include using the same preprocessing code and pipelines during training as those used in serving, A/B comparing any optimized production code used in serving against reference code to ensure comparable results, ensuring at least a subset of the test data is tested directly on the production serving infrastructure to measure any inconsistencies, and routinely retraining models with new production data if there is temporal shift in the data over time.

7.4.1 Machine Learning Classification Metrics

Instead of viewing a model's prediction as either correct or incorrect, it is helpful to break down the errors into the following categories (Table 7.1).

7.4.1.1 Accuracy

A common metric used across machine learning is to compute a model's *accuracy* on a test dataset:

$$accuracy = \frac{number\ of\ correct\ instances}{total\ number\ of\ instances} = \frac{tp + tn}{tp + tn + fp + fn} \tag{7.31}$$

However, accuracy can be a misleading metric, particularly in use cases where the test data is imbalanced. Imbalanced datasets are common in many real world applications, such as those with time series signals like radar presence sensing or gesture sensing where the background class can significantly outnumber the positive classes. Think of a user performing gestures on a wearable watch, mobile phone, or kitchen speaker—even if they are performing one gesture every minute over the course of an hour of use (which is unlikely), there would still be over a 59:1 ratio between nongesture motions and gesture motions being captured by the sensor. Accuracy is a poor metric to use in these scenarios, as a naive model that always outputs the background class regardless

Table 7.1
Model Prediction Error Types and Description

Prediction	Error Type	Description	Example
True positive (TP)	n/a	The model correctly predicts a positive label for a positive instance.	A gesture is correctly predicted as a gesture.
True negative (TN)	n/a	The model correctly predicts a negative label for a negative instance.	The model correctly predicts that no user is present in front of the device.
False positive (FP)	Type 1 error	The model incorrectly predicts a positive label when the true label is negative.	The model incorrectly predicts a gesture was performed when a generic motion was near the sensor.
False negative (FN)	Type 2 error	The model incorrectly predicts a negative label when the true label is positive.	The model incorrectly predicts that no user is present when a user is standing near the device.
Misclassification error (MC)	n/a	The model confuses one positive class with another positive class (only valid in a multiclass classification task).	The model incorrectly predicts a swipe up gesture, when a swipe down gesture was performed.
Multiprediction error (MP)	n/a	The model triggers multiple events when only one event should be triggered.	The model incorrectly predicts two swipe up gestures directly after one another, when the user made a single swipe up gesture.

of the input would still score 98% accuracy if the test dataset had a 59:1 background-to-gesture ratio. This is more formally known as the accuracy paradox, where accuracy is a poor metric for predictive models when classifying in predictive analytics. More rigorous evaluation metrics should therefore be used.

7.4.1.2 F Score, Precision, and Recall

A more robust metric over accuracy is the *F score*, which combines a model's *precision* with a model's *recall*.

Precision is the fraction of correct instances from the detected instances:

$$precision = \frac{tp}{tp + fp} \tag{7.32}$$

Recall is the fraction of correct instances that are successfully detected:

$$recall = \frac{tp}{tp + fn} \tag{7.33}$$

The most common F score is the harmonic mean of the precision and recall, which is known as the F_1 score:

$$F_1 = 2 \bullet \frac{precision \bullet recall}{precision + recall} = \frac{tp}{tp + 0.5(fp + fn)} \tag{7.34}$$

A model with a precision close to 1 indicates the model is precise and has a low number of false positive errors. Conversely, a model with a precision close to 0 indicates the model is imprecise and makes a large number of false positive errors.

A model with a recall close to 1 indicates the model is good at successfully detecting true instances in a test dataset and makes few false negative errors, whereas a model with a recall close to 0 indicates the model fails to detect a large number of true instances in a test dataset and has a large number of false negative errors.

The main benefit of precision and recall is their relationship to each other—it is possible for a bad model to have very good recall but poor precision. Likewise, a bad model could be very precise, but have poor recall and miss most true examples in the test dataset.

For a model to perform correctly, it must have good precision and good recall. The F_1 score combines precision and recall so that an F_1 score close to 1 can be attained only if both recall and precision are also close to 1.

7.4.1.3 F Beta Score

The F_1 score weights precision and recall equally. In use cases where either precision or recall is more important, the F_β score can be used instead, where β reflects the importance of recall over precision.

For instance, a β factor of 2 would weight recall twice as much as precision, with a β of 0.5 weighting precision twice as much as recall:

$$F_\beta = (1+\beta^2)\frac{precision \bullet recall}{(\beta^2\ precision) + recall} = \frac{(1+\beta^2)tp}{(1+\beta^2)tp + \beta^2(fp + fn)} \tag{7.35}$$

7.4.1.4 Multiclass Metrics

In multiclass tasks, the precision and recall can be computed independently for each of the k classes:

$$precision_k = \frac{tp_k}{tp_k + fp_k} \tag{7.36}$$

$$recall_k = \frac{tp_k}{tp_k + fn_k} \tag{7.37}$$

Computing the per-class precision and recall has several advantages. The two most significant are that it becomes obvious if a specific class is underperforming, and computing the per-class results helps mitigate unbalanced class statistics.

Given the per-class precision and recall results, a single F_1 can be computed across all K classes. In some use cases, it may be helpful to weight the score for each class to indicate which class has higher or lower importance to the overall metrics:

$$F_1 = \sum_{k=1}^{K} w_k \frac{tp_k}{tp_k + 0.5\left(fp_k + fn_k\right)} \tag{7.38}$$

where w_k is the weight for the kth class. If each class has an equal weight, then the constant of $\frac{1}{K}$ can be used for all weights.

7.4.1.5 Confusion Matrix

Computing the precision and recall of each independent class can help highlight which of the K classes in the model are performing well and which are underperforming. Precision and recall do not tell the whole story though, such as how many instances positive class i might be confused with positive class j. This is where a confusion matrix can help.

A confusion matrix, also known as an error matrix or matching matrix in unsupervised learning, is a table that allows visualization inspection of the performance of a classification model. Each row in the confusion matrix represents instances in the true class, while each column represents the instances in the predicted class. A perfect classifier would therefore have values along the diagonal of the matrix (where true == predicted) and zeros everywhere else. For nonperfect classifiers, any values not on the diagonal help explain the types of errors made by that model.

7.4.2 Classification Metrics for Time Series Data

Thus far, the metrics described have not considered the temporal nature of the data. Care should be taken when working with sensors like radar that are inherently time series based. Metrics such as accuracy, precision, and recall can all provide misleading values when directly applied to the raw sample rate of the sensor. Applying these metrics at sensor data rates can, at best, give false confidence in data coverage, and at worst hide significant gaps in a machine learning model's performance.

7.4.2.1 Event Based Time Series Metrics

Instead of directly applying these metrics to the raw sensor data, it can be helpful to estimate key metrics based on *events* that occur in the time series data, not on the counts of the raw time series samples themselves.

For example, in a gesture recognition context, metrics such as precision and recall should be computed based on the number of discrete gesture motions performed by users, not by the number of raw data samples that observe these motions.

Imagine you were building a radar-based gesture-controlled light bulb that would turn on or off if a user performed a circle motion with their hand within range of the sensor. To test the performance of your algorithms, you collect a test dataset consisting of 10 unique users each performing 10 examples of the circle gesture at different locations near the sensor, giving a test set of 100 gesture instances. In addition to the 100 gesture examples, you ask each user to perform various nongesture activities in front of the sensor for 10 minutes, resulting in 100 minutes of background recordings. Let's assume that the average duration of each user's circle gesture is 1 second and the radar sensor has a pulse repetition frequency of 2,000 Hz. Given this, it would result in a test dataset consisting of:

- 100 unique gesture instances;
- 100 * 2,000 = 200,000 samples of radar data with positive gesture data;
- 100 * 60 * 2,000 = 12,000,000 samples of radar data with nongesture motions.

It's clear from these numbers that evaluating the test dataset on a sample-by-sample bias significantly skews the performance of the model. Instead, a more robust metric is to compute the detection of the gesture events based on the unique number of gesture instances (i.e., 100).

7.4.2.2 Latency

An additional metric that is important for interactive systems is latency. Latency can be defined as the time delay between an action being performed by a user and the system reacting to that action. For example, the delay between a gesture being made by a user and the detection of that gesture triggering some music to play, or the delay between a user walking into the room and being detected by a radar and the light being triggered to turn on by a smart-home system.

Latency requirements will vary based on specific use cases, and could range from ~20 ms for time-sensitive systems (e.g., professional musical instruments) to ~100 ms for interactive controls (e.g., touchscreen latency), to several seconds for other systems (e.g., dimming or resetting the screen on a display when a user walks away).

Use cases with stricter latency requirements are more challenging from an algorithm design perspective, as they require the algorithm to trigger a detection event as soon as it is observed without the luxury of integrating over more time to increase the confidence in a detection.

For example, when detecting gestures, there are instances where it is easier to infer the true label of a motion only when the full motion has been completed. This is because the start and mid points of some gestures can look very similar to other random background motions, and it can only be at the end of the motion that there is enough information to distinguish true gestures from harder background motions.

Detection rates can be increased and false positive errors decreased by analyzing more of the motion at the end of a potential gesture to increase detection confidence before triggering a detection event. However, while delaying longer can increase detection accuracy, it can significantly reduce the responsiveness of an interactive system, which can ultimately frustrate and fatigue users and quickly remove all value from an interactive product. There is therefore an important tradeoff between system robustness and system responsiveness.

Latency is an important factor when measuring algorithm performance, as the algorithm with the top detection rate may not necessarily be the "best" algorithm if its detection latency is too high.

When measuring a latency, it is important to take into account the full system latency, not just the algorithm detection latency. When measured in a holistic system, the algorithm latency might consume only a small percentage of the total latency observed by the user. Therefore, the latency target for any interactive system should reflect the total latency observed by the user and not just the latency of a given algorithm. This latency could include the sample rate of the sensor, the update rate of the algorithm, the detection latency of the algorithm, the software delay of propagating detection events throughout an operating system or application, the loading time of any consuming applications, and the refresh rate of screen and/or audio system.

7.5 The Future of Radar Machine Learning

There have been significant advances in radar technology in recent years, resulting in a series of innovative new products in wearables and personal computing, alongside fundamental progress in radar-based research and development.

While radar technology is almost a century old, it is clear that we are only beginning to see how this exciting technology can be applied to personal computing and human-machine interaction.

7.5.1 What's Next?

So what's next in the future of radar-based machine learning?

A good proxy for predicting the next five-plus years of advancements in radar technology is to look at the major trends in more mature machine learning fields, such as computer vision or natural language processing. This includes moving from supervised learning to self-supervised learning, improving model architectures with meta-learning, and providing better baselines through open source datasets and libraries.

7.5.2 Self Supervised Learning

Recent advances in fields like computer vision or natural language processing have shifted toward leveraging very large unlabeled datasets to pretrain incredibly large models to achieve state-of-the-art results over traditional supervised learning techniques [23, 24].

These self-supervised techniques use the data itself as the label, which removes the significant bottleneck of supervised datasets. Small supervised datasets are still typically required to test the efficacy of self-supervised techniques, but they can be orders of magnitude smaller than traditional supervised datasets. This makes these techniques particularly advantageous for novel sensors, such as radar, which lack large supervised datasets for research or production use cases.

7.5.3 Meta Learning

In the early days of machine learning, engineers would hand-craft the weights of the network directly. Then came a few decades of engineers carefully hand-designing the features that would be fed into the machine learning algorithms, such as hand-designed computer vision feature techniques like histogram of gradients [12], scale-invariant feature transform (SIFT) [13], or speeded-up robust features (SURF) [14]. This was followed by the era of deep learning, where much larger and deeper neural networks would directly learn the features from the direct sensor inputs [8], with machine learning engineers carefully crafting the architectures of these networks to maximize model performance.

The most recent trend is to have a machine learning algorithm attempt to learn the best architecture for a given problem instead of a human engineer hand-designing the architecture. In other words, have a machine learning model train many machine learning models and *learn* the optimal architecture or hyperparameters for a specific problem. This technique is known as meta-learning, learning-to-learn, neural architecture search (NAS), or auto ML.

There are many techniques behind auto ML, but they all fundamentally relate to some form of intelligent search through a predefined search space defined by a human designer. Custom architectures found by various NAS techniques have resulted in state-of-the-art performance across numerous machine learning benchmarks [25, 26].

While auto ML techniques are automating and accelerating the architecture design stage of the machine learning process, care should be taken when using these techniques, particularly for novel sensors such as radar. The best auto ML techniques as of today can provide state-of-the-art performance on well-established tasks, such as object detection in computer vision. However, these techniques have been shown to fail quickly when applied to novel sensors or tasks outside the original domain. In other words, one cannot simply take the state-of-the-art auto ML search techniques for computer vision images and naively apply it to radar data and instantly outperform an architecture design by a human expert—at least not today.

Nonetheless, these auto ML techniques are rapidly improving the state of the art for machine learning architecture design, and we can expect to see similar progress in radar-based machine learning architecture refinement as these techniques improve and generalize across domains.

7.5.4 Sensor Fusion

Thus far, our focus in this chapter has been on using radar as the primary sensor. In a real-world product scenarios, however, radar would most likely be just one sensor in a suite of sensors. When combined with other sensors, such as RGB cameras, inertial sensors, or other complementary sensors, radar has two fundamental areas of impact.

The first is where the radar is used as a low-power trigger for other sensors. One example might be where an always-on radar operating in a low-power mode is used to detect the potential presence of a user, which then triggers a camera to warm up and confirm the presence of a specific user. In this sensor fusion scenario, we can take advantage of the low-power properties of the radar to help improve the efficiency of more power-hungry sensors, such as high-resolution cameras.

The second area where radar can have a significant impact is via fusion with other complimentary sensors (e.g., combining radar and RGB cameras to unlock new use cases that neither a radar nor a camera could support independently). Radar is a particularly exciting sensor for sensor fusion, as it provides fundamental information that may not be captured by other sensors, such as cameras or other wearables like inertial sensors.

7.5.5 Radar Standards, Libraries, and Datasets

Another fundamental pillar that underpins the incredible success of modern machine learning algorithms in the fields of computer vision and natural language processing are the well-established standards, libraries, and datasets required to rapidly advance these fields.

The impact of open-source libraries such as OpenCV [27] or datasets such as ImageNet [9] have had on both the academic and production communities should not be underestimated. These libraries and datasets allow researchers to quickly replicate prior work or easily benchmark their work on identical datasets.

Establishing similar standards, libraries, and datasets for radar will play a fundamental role in advancing the field of radar-based machine learning.

7.6 Conclusion

The fields of radar and machine learning have long-established, but mostly independent, histories. It's only been recently that the best aspects from both fields have started to be combined to create a new category of exciting sensing solutions and use cases.

There are myriad opportunities for how machine learning can be combined with radar to develop enhanced detection algorithms; gesture and activity recognition algorithms; sensor fusion with radar, cameras, and/or other sensors; health and biometrics applications; privacy persevering use cases where cameras are not suitable; and beyond.

References

[1] Bishop, C.M., *Pattern Recognition and Machine Learning,* New York: Springer, 2006.

[2] Russell, S. J., *Artificial Intelligence: A Modern Approach,* Second Edition, Upper Saddle River, NJ: Prentice Hall, 2002.

[3] Goodfellow, I., et al., *Deep Learning,* Cambridge, MA: The MIT Press, 2016.

[4] Mitchell, T., *Machine Learning,* New York: McGraw-Hill, 1997.

[5] Rosenblatt, F., "The Perceptron: A Probabilistic Model for Information Storage and Organization in the Brain," Cornell Aeronautical Laboratory, *Psychological Review,* Vol. 65, No. 6, 1958, pp. 386–408.

[6] Minsky, M., and S. Papert, *Perceptrons: An Introduction to Computational Geometry,* Cambridge, MA: The MIT Press, 1969.

[7] LeCun, Y., et al., "Backpropagation Applied to Handwritten Zip Code Recognition," *Neural Computation,* Vol. 1, 1989, pp. 541–551.

[8] Gillian, N., "Gesture Recognition for Musician Computer Interaction," Ph.D. thesis, 2011.

[9] Wang, S., et al., "Interacting with Soli: Exploring Fine-Grained Dynamic Gesture Recognition in the Radio-Frequency Spectrum," *Proceedings of the 29th Annual Symposium on User Interface Software and Technology*, 2016, pp. 851–860.

[10] Hayashi, E., et al., "RadarNet: Efficient Gesture Recognition Technique Utilizing a Miniature Radar Sensor," *Proceedings of the 2021 CHI Conference on Human Factors in Computing Systems*, 2021, pp. 1–14.

[11] Pan, et al., "OpenRadar," GitHub, 2019, https://github.com/presenseradar/openradar (last accessed May 2022).

[12] Deng, J., et al., "Imagenet: A Large-Scale Hierarchical Image Database," *IEEE Conference on Computer Vision and Pattern Recognition*, 2009, pp. 248–255.

[13] Lowe, D. G., "Distinctive Image Features from Scale-invariant Keypoints," *International Journal of Computer Vision*, 60(2), 2004, pp. 91–110.

[14] Bay, H., et al., "Speeded-Up Robust Features (SURF)," *Computer Vision and Image Understanding*, Vol. 110, No. 3, 2008, pp. 346–359.

[15] Howard, A., et al., "Searching for Mobilenetv3," *Proceedings of the IEEE/CVF International Conference on Computer Vision*, 2019.

[16] Cortes, C., and V. Vapnik, "Support-Vector Networks," *Mach. Learn.*, 20, 1995, pp. 273–297, https://doi.org/10.1007/BF00994018 (last accessed May 2022).

[17] Breiman, L., "Random Forests," *Mach. Learn.*, 45, 2001, pp. 5–32, https://doi.org/10.1023/A:1010933404324 (last accessed May 2022).

[18] Dalal, N., and B. Triggs, "Histograms of Oriented Gradients for Human Detection," *IEEE Computer Society Conference on Computer Vision and Pattern Recognition*, Vol. 1, 2005, pp. 886–893.

[19] Lien, J., et al., "Soli: Ubiquitous Gesture Sensing with Millimeter Wave Radar," *ACM Transactions on Graphics*, Vol. 35, Issue 4, No. 142, 2016, pp. 1–19.

[20] Gillian, N., et al., "The Gesture Recognition Toolkit," *Journal of Machine Learning Research*, Vol. 15, Issue 1, 2014, pp. 3483–3487.

[21] Gillian, N., et al., "Recognition of Multivariate Temporal Musical Gestures Using N-Dimensional Dynamic Time Warping," *Proceedings of the International Conference on New Interfaces for Musical Expression*, 2011, pp. 337–342.

[22] Alom, M. Z., et al., "The History Began from Alexnet: A Comprehensive Survey on Deep Learning Approaches," arXiv preprint arXiv:1803.01164, 2018.

[23] Brown, T. B., et al., "Language Models Are Few-Shot Learners," arXiv preprint arXiv:2005.14165, 2020.

[24] Chen, T., et al., "Big Self-Supervised Models are Strong Semi-Supervised Learners," arXiv preprint arXiv:2006.10029, 2020.

[25] Schmidhuber, J., "Deep Learning in Neural Networks: An Overview," *Neural Networks*, Vol. 61, 2015, pp. 85–117.

[26] Zoph, B., et al., "Learning Transferable Architectures for Scalable Image Recognition," *Proceedings of the IEEE Conference on Computer Vision and Pattern Recognition*, 2018, pp. 8697–8710.

[27] Bradski, G., "The OpenCV Library, Dr. Dobbs," *Journal of Software Tools*, 2000.

8

UX Design and Applications

Leonardo Giusti, Lauren Bedal, Carsten Schwesig, Eiji Hayashi, and Ivan Poupyrev

8.1 Overview

Human-machine interfaces rely on a necessary interplay between user input and system output. Input is a key component of interaction; it allows a system to understand human intent and orchestrate complex tasks to satisfy user needs. Some of the first input methods date back to the pen interface, developed by Ivan Southerland [1]; the computer mouse by Engelbart; the "put it there" gesture interface by Bolt, PHIGS, and GKS [2]; and the Nintendo Power Glove [3]. This research was revolutionary for the time and shaped the graphical user interfaces we use today to interact with technology.

As the context and form of computing change, so do input technologies. This has been particularly seen in the change from desktop to mobile computing, where touch-screen technologies have given us constant access to information anywhere we are.

As we shift from mobile to ambient computing, alternative input methods, such as voice [4], gaze [5], gestures, and even biosensing [6], are gaining in popularity. Voice input is now widely based on use of personal assistants on mobile phones (e.g., Google Assistant and Apple's Siri [7]) and smart speakers. Gestural interactions have also grown in popularity, such as depth cameras for human posture recognition from 3-D hand pose estimation using depth images [8]. In addition, noninvasive functional neuroimaging techniques, EEG in particular (such as NeuroSky, Emotiv, and OpenBCI), have been studied as an additional form of continuous data input for ambient computing devices [9].

We have focused on the invention of a new interaction language using radar for gesture recognition [10], proximity detection, tracking people, body language, and even biofeedback signals. Radar technology is an attractive candidate as an input technology for ambient computing where previous direct touch interaction is difficult, if not impossible. Radar is lower cost, smaller, and more privacy preserving than camera-based solutions [11, 12]. Due to the small form factor, low-power consumption and low cost of RF-based sensors, we can envision use of radar technology in wearable and mobile devices, smart appliances, interactive robots, game controls, large-scale interactive physical environments, and more.

This chapter provides a deeper insight into designing with radar as an input method for human-computer interaction. In this chapter, we will cover the fundamentals and characteristics of radar for designers, an interaction language based on radar, and finally, a set of case studies of consumer products invented at Google ATAP.

8.2 Understanding Radar for Human-Computer Interaction

As designers, it is critical to understand the fundamental principles of radar sensing in order to properly design interactions. Radar works by using modulated electromagnetic waves that are emitted toward a moving or static target that scatters the transmitted radiation. Some portion of energy is redirected back toward the radar, where it is intercepted by the receiving antenna as an example shown in Figure 8.1. The time delay, phase or frequency shift, and amplitude attenuation contains rich information about the target's properties,

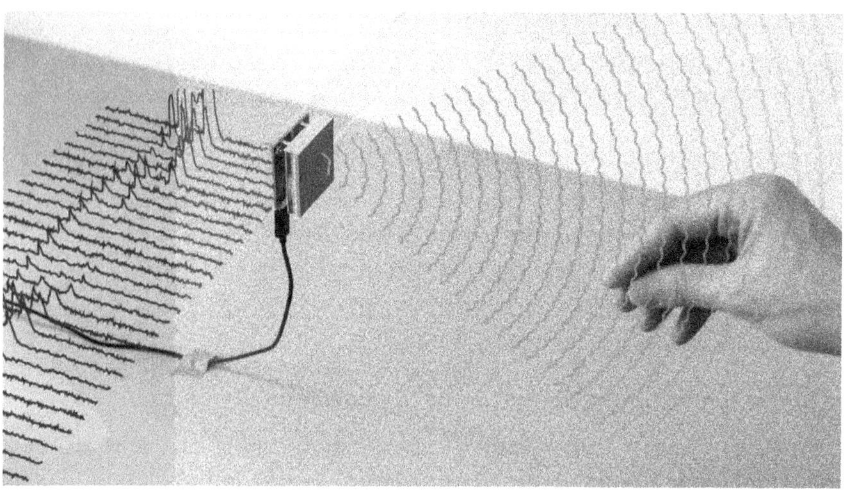

Figure 8.1 Radar capturing motion of the human hand.

such as distance, velocity, size, shape, surface smoothness, material, and orientation, among others. Thus, these properties may be extracted and estimated by appropriately processing the received signal.

The goal of any radar system design is to optimize radar functional performance for the specified application, such as gesture tracking, multiple person tracking, and microgestures. It is important that the entire radar system, inclusive of hardware, signal processing techniques, and radar control software and algorithms, is strongly interconnected and cannot be specified without tracking into account specifics of a front-end application.

Radar can be applied to a wide range of scales, ranging from hundreds of miles to submillimeter gesture detection. However, for the purposes of using radar for human-computer interaction, we have focused on the scales of human motion, from small finger movements to room-scale presence and motion (Figure 8.2).

There are a variety of existing gesture sensing approaches. Some rely on skeletal tracking of specific poses of the body, such as the Microsoft Kinect. Other gestural systems such as the Hololens rely on explicit estimations of a hand pose as a "mode switch" prior to recognizing gestures [13]. Radar provides a fundamentally different approach to gesture recognition by exploiting the temporal accuracy of radar and extracting and recognizing motion signatures. In addition to exploiting the temporal accuracy of radar, gesture recognition can be designed to (1) maintain the high throughput of the sensor to minimize latency; (2) exploit advantages of using multiple antennas to maximize SNR and improve recognition accuracy; (3) provide both discrete and continuous predictions; and (4) be computationally efficient to work on miniature, low-power devices (e.g., smart watches).

In order to establish meaningful design principles, we first have to avoid simplistic mental models that are based on familiar existing applications:

- *Radar is not just a speed sensor:* speed guns, such as the ones used by traffic police or in baseball games (Figure 8.3), suggest that pointing a radar sensor at a moving object would give an accurate, absolute measure of the speed of an object. In reality, the data generated by a radar sensor

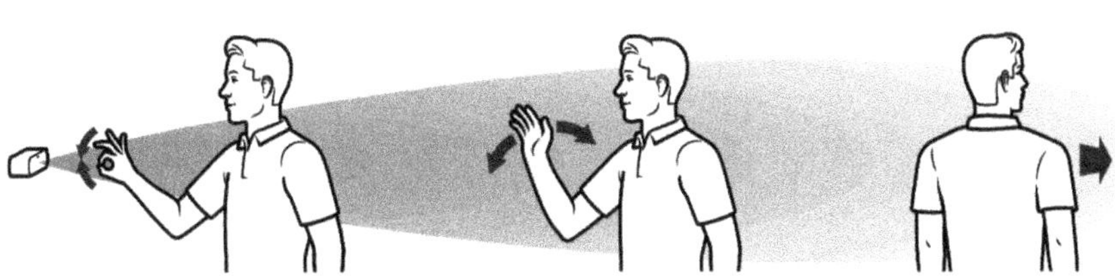

Figure 8.2 Radar can detect different scales of movement from microgestures to full body movement.

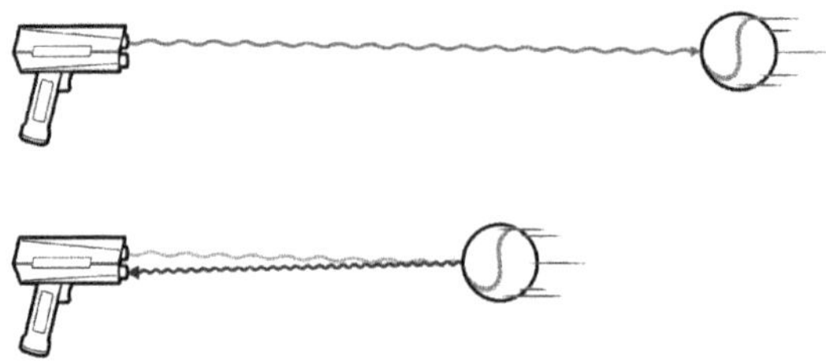

Figure 8.3 Radio waves are sent out into the environment, hit a target, and bounce back.

is much more complex and can detect, for example, multiple motions within its environment.

- *Radar is not just a 2-D object sensor:* in more traditional applications, naval radar sensors display an interface showing moving objects within a circular range around the sensor, as shown in Figure 8.4. In reality, this type of display would be difficult to achieve in a human-scale environment, where objects are not as clearly distinguished from the space around them as ships are in an ocean.
- *Radar is not a 3-D camera:* although radar is used to map topographic features, we quickly realized that it is counterproductive to think of a human-scale radar sensor as capable of producing real-time 3-D scans of its environment. The skeleton models available through Kinect-like sensors would not play to the strengths of radar sensing, as illustrated in Figure 8.5.

To understand the nature of the data generated by radar sensing at the human scale, engineers and designers are working together to explore how radar might be used for interactions. For example, Figure 8.6 is the result of mapping a camera shutter to a threshold range detected by the radar. It clearly illustrates the effect a hand target's shape and size has on the range detected by the sensor.

Range Doppler visualization is also very helpful in establishing the understanding of how radar works for interaction. It detects motion components in the sensor field and maps motion directionality to the y-axis and motion velocity to the x-axis. Experimenting with hand gestures seen through the range Doppler visualization helped visualize radar's unique capabilities in detecting multiple motion components at a very high refresh rate. Although the sensor does not see hand gestures as 3-D images, repeated gestures are recognizable in how they appear in the range Doppler visualization.

Our experiments and early explorations led to some foundational principles for radar-based human computer interaction:

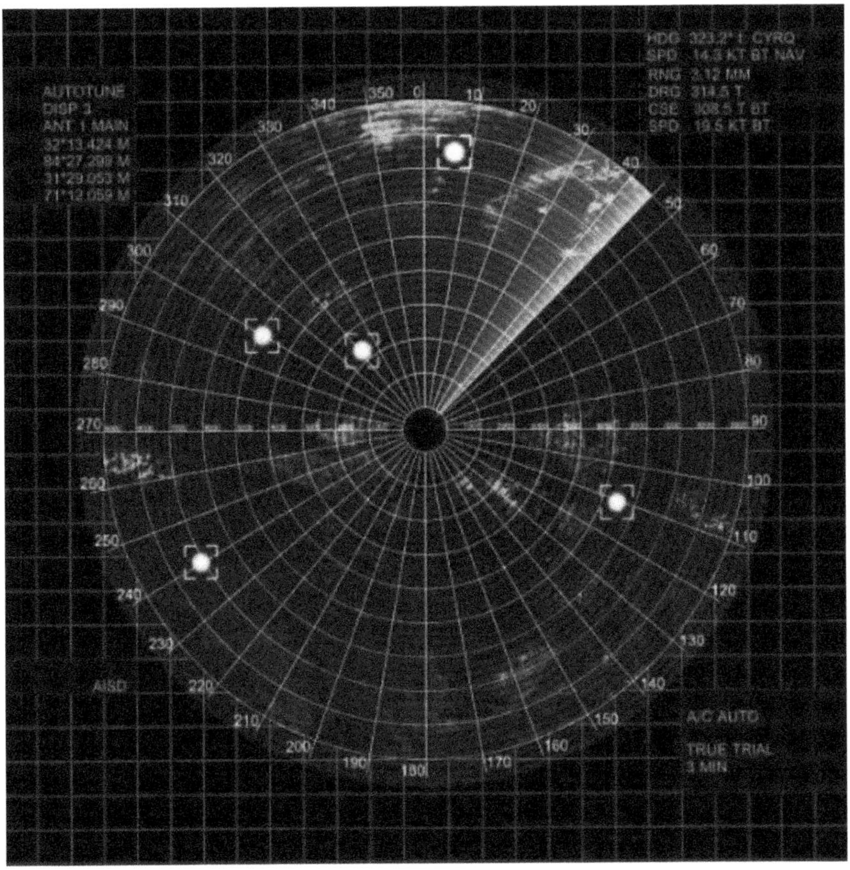

Figure 8.4 Traditional radar systems detect large moving objects.

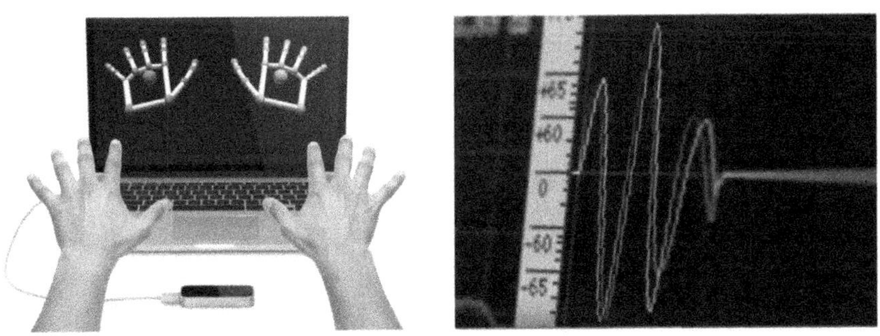

Figure 8.5 Unlike skeletal tracking with the Kinect, radio waves sense qualities of motion.

- Focus on simple interactions with a clear motion signature: radar captures motion profiles and temporal variations, not images or 3-D shapes.

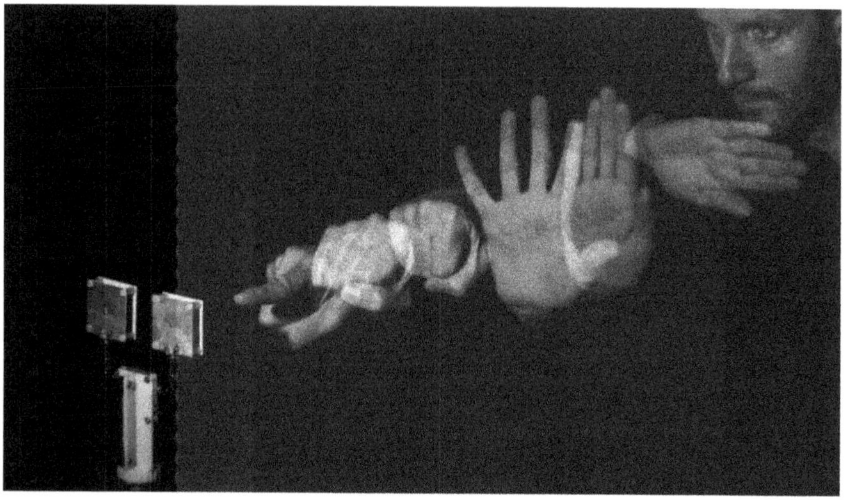

Figure 8.6 Early design studies clearly illustrate the effect that a hand target's shape and size has on the range detected by the sensor.

In designing gestures for radar detection, specific poses will be difficult or impossible to detect. Instead of relying on poses, gestures should be defined by their distance, velocity, and amplitude relative to a radar sensor.

- Design gestures relative to the sensor: in the case of hand gestures, the hand's shape, size, and distance from the sensor have an impact on how the gesture will be interpreted. This can lead to unpredictable results if the hand's location in relation to the sensor is not taken into account, but the sensor location can provide a useful frame of reference for various types of gestures—microgestures can be detected close to a sensor, hand and arm movements farther away, and other implicit signals like human presence even further.
- Take advantage of the unique qualities of radar:
  - *Range tracking:* radar is especially sensitive to movements perpendicular to the sensor plane (radial). Movements that are parallel to the sensor plane (tangential) are much more difficult to detect.
  - *Update rate:* radar sensing is very fast (>1,000Hz) in comparison to photography. It can reliably detect changes in its environment much faster than the 30 frames per second typical in a video camera.
  - *Private:* the diffusion of products with advanced sensing capabilities in personal spaces, such as our bedrooms, living rooms, or workplaces, makes privacy a key factor for their wide adoption.

- *Small:* gesture recognition techniques should have a small footprint in order to be embedded in a variety of objects without compromising their form factor or aesthetic.
- *Invisible:* such techniques should disappear behind surfaces, without requiring openings or other modifications.
- *Robust:* In the context of our homes, environmental conditions change constantly (e.g., light can be dimmed), and movements that occur naturally might be easily mistaken for gestures. The RF signal used in radar sensors is not affected by ambient light or noise, making the system more robust against environmental changes compared to cameras and microphones. Finally, radar's sensitivity to motion regardless of distance enabled gesture recognition in both near and far fields.

8.3 A New Interaction Language for Radar Technology

As discussed in the previous section, understanding radar for human-machine interaction, radar provides robust, private, and continuous sensing at different scales.

In this section, we will present a library of interaction primitives and a framework to organize, scale, and apply them to a variety of products and applications.

Radar technology offers the opportunity to define a new interaction language that extend existing interaction techniques in two main ways:

1. Complementing existing *explicit* input techniques, such as voice and touch, with reliable, efficient, and affordable gestures detection;
2. Extending the interaction cycle beyond direct interaction, including the possibility of capturing indirect (*implicit*) interactions between people and devices.

8.3.1 Explicit Interactions: Gestures

Explicit interactions are methods of interacting with a system that involves a direct intention from the user to interact with a system. Touch gestures we all use today with mobile phones are a good example of explicit interactions. For example, when swipe is detected, the system understands to advance to the next image in a carousel.

From early research work to more recent product launches such as Microsoft Kinect, most of the work on gesture recognition techniques has focused on image-based sensors for specific tasks, such as gaming or 3-D manipulation. Despite the quick advancement in the human pose estimation, the limitations, such as a narrow field of view, challenges in tracking quick motions, and strong privacy concerns around keeping cameras always-on in private spaces, have prevented the penetration of these camera-based gestural interactions in more common places, such as a home environment.

Radar technology can provide efficient solutions for explicit and gestural recognition for consumer products. Unlike image-based gesture sensors, radar directly captures gesture motion profiles and temporal variations in scattering center properties as motion signatures for gesture recognition. This approach reduces the computational load, making real-time radar-based gesture recognition feasible even on low-power devices such as smart watches and IoT.

Devices for ambient computing can benefit from gestural interactions in many ways. For example, in the context of our homes, the user's attention is often split between multiple tasks at the time, and interactions with products are part of a larger choreography of activities such as cooking, chatting with a friend, or getting ready to leave the house. These interactions do not always happen near devices, making touch-based input modality, which requires users to come close to devices or to always keep devices in reach, less attractive. In this context, gestures provide a valuable alternative: they provide the opportunity to define an input method that doesn't require precise hand-eye coordination, minimizing the user's cognitive load and allowing the user to access information and services while staying engaged with their primary tasks (e.g., cooking). Therefore, gesture interactions with radar can be:

- *Immediate*: The gestures recognition system can run continuously, and it is ready anytime the user wants to initiate the interaction. It does not require a "wake-up" gesture or trigger. The value of gesture-based interactions in consumer electronics is in their immediacy: any friction, such as the need to wake up the device, will impact the usefulness of gestural interactions in these contexts.
- *Eyes free*: Gestures can be performed anywhere in a sensing area, allowing the user to perform gestures with minimal hand-eye coordination.
- *Reusable*: The high-level mapping of movement primitives (e.g., tap to take action or swipe to move content) can support a variety of use cases based on context (e.g., sweep to skip track or dismiss an alarm). An illustration of music control is shown in Figure 8.7.

In conclusion, radar technology offers the unique opportunity to commercially deploy gestures in a variety of contexts and products, overcoming some of the limitations of different types of sensors such as cameras. While

Figure 8.7 A quick swipe of the hand over the Pixel 4 can advance to the next/previous track.

gestures have been mainly developed for very specific tasks and applications, radar technologies allow us to create ubiquitous products and experiences in our everyday life.

8.3.2 Implicit Interactions: Anticipating Users' Behaviors

Implicit interactions employ natural, everyday movements of the user as a mechanism to orchestrate a device response. Understanding if a user is nearby a device is probably the simplest type of implicit interaction. For example, when presence is detected by an automatic door, it opens. There are several examples of RF sensing systems used to detect people's presence, including counting people and locating multiple users. Radar has also been used to detect more complex implicit behaviors, such as the intention of the user to grab a device (see the use case on the Pixel 4 in Section 8.4) or even vital signals related to heartbeat and respiration for sleep tracking (see the use case on the Nest Hub in Section 8.4) or other passive health care–related features.

Implicit interactions differ from traditional interaction cycles where traditional digital products, such as gaming consoles, desktop computers, laptops, mobile phones, and smart TVs were either on or off, and an interaction session usually started by turning on the device. An interaction session was maintained through a set of explicit commands with a device.

Expanding our interaction language from explicit interactions to include implicit interactions is especially useful in an ambient computing context, where a growing number of always-on digital devices have populated our living environments: intelligent thermostat, voice-activated speakers, smart air-purifier, home displays, and so on.

Often, these devices have minimal context of user intent in order to orchestrate information and manage transitions, resulting in less than ideal experience. For example, the Nest thermostat always lights up when you walk past it, even if you have no intention of interacting with it; while hot words that wake up a microphone create frictions, making conversations with the device less natural.

Implicit interactions can address some of these challenges by understanding a person's location and level of engagement with a device as a way to modulate the behavior of the device, as well as continuously adapt to the ongoing user activity. Implicit interactions activate input modalities only when needed (such as a mic), changing power and privacy settings based on users' behaviors, and adapting the presentation of information to changes in personal and social context. For example, a Nest thermostat monitors room occupancy in the background and gets ready for interaction when people approach it; an Assistant-enabled speaker is always listening for the "OK Google" hot words to initiate the conversation. Figure 8.8 demonstrates that a notification can be expanded when the user is looking in that direction.

In summary, radar technology offers the opportunity to define a new interaction language that extends existing interaction techniques by complementing existing explicit input techniques, such as voice and touch, with reliable, efficient, and affordable gestures detection, and also adds the possibility of capturing indirect (implicit) interactions between people and devices.

Figure 8.8 Implicit cues like a glance can expand notifications.

8.3.3 Movement Primitives

For interaction designers, building design systems and components is a normal part of the design process. However, designing input methods is less common in day-to-day interaction design work. When designing new interaction languages with radar, one challenge is to formalize the language for an experience when the system input is inherently spatial. This requires (1) defining the atomic units of movement detected by radar, and (2) defining the logic in how these units work together with front-end applications to define the overall user experience.

We introduce the concept of movement primitives as a way to build interactive systems for computing with any enabling motion-sensing technology, including radar. Movement primitives provide a common language for designers and developers, and can be employed throughout the research and product development process.

Movement primitives are building blocks of spatial interaction that consist of two components, a movement signature plus system response. First, a movement signature is a pattern of movement derived by the spatial dimensions of the radar signal, such as proximity, pathways, velocity, amplitude, and so on. These signals can be implicit behaviors (such as proximity and body language) or explicit (such as direct gestures, voice, or touch). Each movement primitive can have specific outputs based on use cases and technological feasibility. A system response consists of a digital or physical characteristic of a computational system.

Movement primitive examples are presented in Table 8.1.

Movement primitives are a useful tool throughout both research and product development stages, as they aid in formalizing the interaction language used for a product as well as the underlying software API. When inventing or implementing a new movement primitive for a consumer product, we typically capture both design and technical details as follows:

- *Definition*: A description of the movement primitive in both user-facing and technical terms. For our work, this typically includes a user-facing definition of the movement, as well as a detailed description covering key aspects of motion such as anatomical change, speed, amplitude, distance, and orientation relative to the sensor.
- *Use cases*: A list of use cases that a particular movement primitive will be applied to. This will help guide the definition of UX and technical requirements, testing scenarios, and output parameters, or APIs.
- *UX requirements*: A list outlining the intended behavior of the system, from a user point of view. This includes how, where, and when this primitive is detected, as well as certain scenarios or contexts in which it

Table 8.1
Example Movement Primitives and Their Corresponding System Responses

Movement Primitive	System Response
Implicit	
Presence	To wake, to activate, to turn on/off
Pass	To wake, to activate, to turn on/off
Sit	To wake, to activate, to turn on/off
Reach	To reveal, to signal, to peek, to grow/expand
Approach	To reveal, to signal, to peek, to grow
Turn	To reveal, to signal, to peek
Explicit	
Swipe	To navigate
Tap	To select, to confirm, to take action
Circle trace	To adjust, to modify, to fine tune
Press (virtual button)	To select, to confirm, to take action
Slide (virtual slider)	To adjust, to modify, to fine tune
Dial (virtual dial)	To adjust, to modify, to fine tune

does or does not work. The UX requirements for the movement primitives are largely driven by the use case applications these primitives will be used for.

- *Technical requirement*: A list of requirements that translates UX requirements into quantifiable metrics, including details on latency, false positives, recall, and precision. This is largely driven by use case applications; separate metrics may be outlined per use case.
- *Testing scenarios*: A list of test case scenarios that are used for algorithmic evaluation. This should include key observational and control variables based on variations of movement, environmental conditions, and user types.
- *Output parameters/APIs*: A list of output parameters that will be exposed to the developers and will help guide front-end applications function. For example, a swipe gesture may have an output parameter to determine if it's a fast or slow swipe. And implicit primitives like presence may have an output parameter to describe not only the distance, but also the orientation relative to the sensor.

8.4 Use Cases

Let's discuss a few examples of how movement primitives can be used to compose use cases for consumer products. These consumer products integrate the 60-GHz Soli [14] radar systems.

Virtual Tools, Microgestures (2016)

Much of the early work using the Soli radar technology focused on capturing the complexities of the human hand through microgestures: hand-scale finger gestures performed in close proximity to the sensor. These micro gestures exploit the speed and precision of the sensor and avoid the fatigue and social awkwardness associated with the arm- and body-scale gesture interactions enabled by popular camera-based gesture tracking techniques.

When designing microgestures, we created a set of interaction primitives that recognized subtle movements of the hand. This interaction language was a granular reflection of the fidelity of the technology and provided a mental model for users to utilize.

1. *Virtual button:* pressing index finger and a thumb as illustrated in Figure 8.9.
2. *Virtual slider:* moving a thumb along the side of the index finger of the same hand as shown in Figure 8.10.
3. *Virtual dial:* an invisible dial operating between thumb and index finger as illustrated in Figure 8.11.

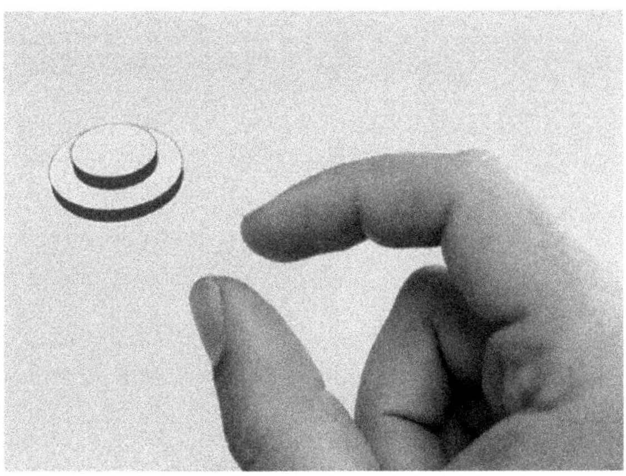

Figure 8.9 Illustration of a virtual button.

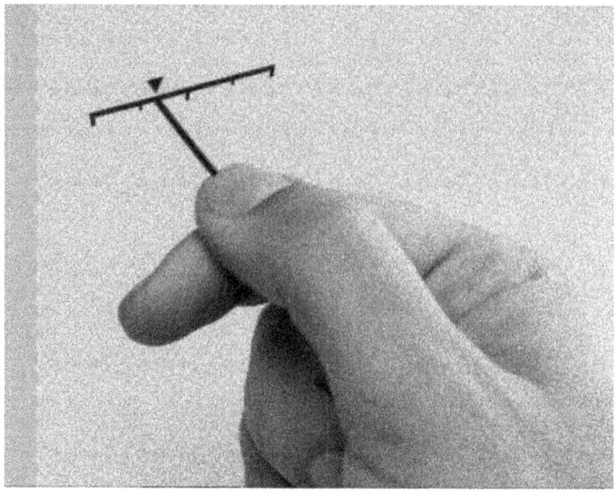

Figure 8.10 Illustration of a virtual slider.

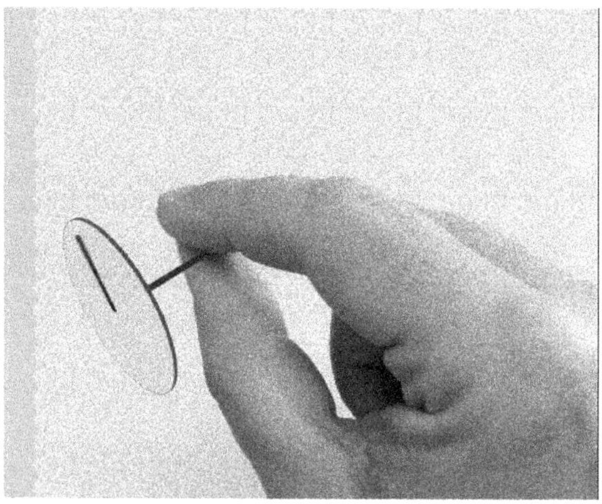

Figure 8.11 Illustration of a virtual knob.

Pixel 4 (2019)

The Soli radar, platform, and interaction model were introduced on the Pixel 4 as motion sense. The sensor is located at the top of the phone (Figure 8.12), creating an interactive hemisphere that can sense and understand movements around the phone. Pixel 4 focused on creating experiences that felt truly in step with the user's daily lives by accelerating and extending controls of common interactions with their phones. It used the following movement primitives:

1. Presence to wake up the display.

Figure 8.12 The Soli radar chip was integrated into the top bezel of the Pixel 4 phone.

2. Reach to reveal the lock screen (and prime cameras for face authentication).
3. Swipe to skip songs or snooze alarms.
4. Tap to pause/resume media.

Nest Thermostat (2020)

The Nest Thermostat, released in 2020, features Soli technology. Soli is elegantly integrated behind the mirror display, so that it can turn itself down to save energy when you leave the house. Since the radar signal penetrates through the plastic and glass, the enclosure is seamless without the need for an aperture, as shown in Figures 8.13 and 8.14.

The Nest Thermostat uses Soli radar to detect motion around the display. It also helps power Nest's home/away assist feature, which uses Soli radar along with GPS to turn down the temperature if users are away and turn it back up when it detects that they are coming home.

Google Nest Hub (2021)

Motion sense was also introduced in 2021 in the Google Nest Hub, Second Edition. This device used a Soli radar embedded in the top bezel of the display as shown in Figure 8.15, creating a sensing cone 1.5m in front of the display. With this sensing area, a variety of use cases are enabled, including gesture controls and sleep detection.

Figure 8.13 Nest Thermostat with Soli radar technology.

Google Nest Hub, Second Edition, introduced two gestures: an omniswipe, meaning a swipe gesture in any direction over the display to dismiss interruptions, as well as a tap gesture to select content. The gestures are organized spatially and mapped to general intents: gestures performed on the plane parallel to the sensor (horizontal plane) such as "swipe up, down, left, and right" are mapped to "navigational" intents (e.g., change track, navigate a list of cards, dismiss an item); gestures performed perpendicularly to the sensor, such as "tap," are mapped to "selection" intents (e.g., pause music, initiate a timer), as shown in Figure 8.16. Gesture/intent mappings correspond to existing mental models for touch-based interactions with tech and are shared cross culturally. The following movement primitives are used:

1. Presence to detect sleep.
2. Swipe to skip songs or snooze alarms.
3. Tap to pause/resume media.

Summary

Human-machine interfaces rely on a necessary interplay between user input and system output. As we shift from mobile to ambient computing, alternative input methods, such as voice [4], gaze [5], gestures, and even biosensing [6],

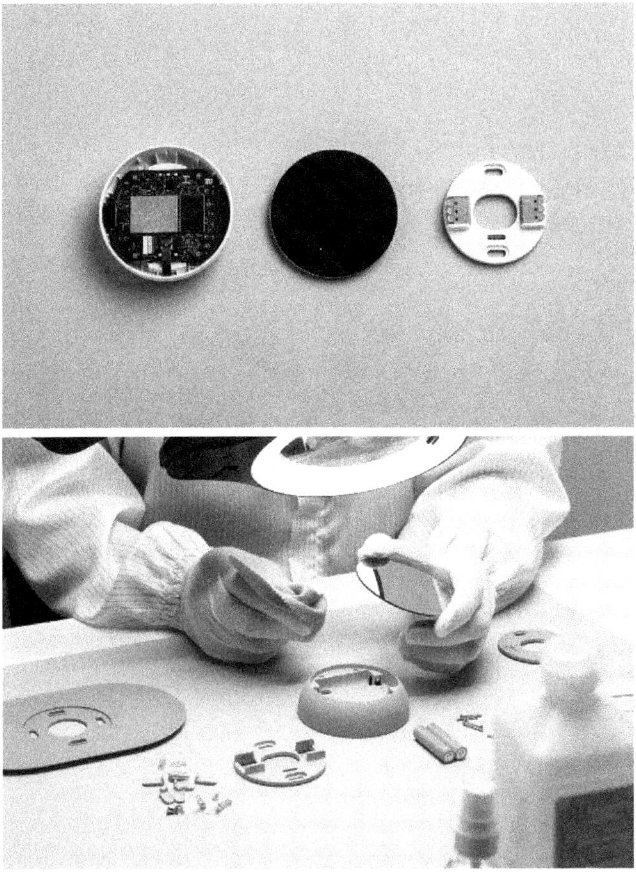

Figure 8.14 Radar penetrates through the plastic and glass, creating a seamless design.

are gaining popularity. This requires designers to understand the fundamental principles of how to design interactions with new types of inputs that understand spatial movement.

For human-machine interaction, radar provides robust, private, and continuous sensing at different scales, both implicit and explicit sensing. Explicit interactions are methods of interacting with a system that involves a direct intention from the user to interact with a system. On the other hand, implicit interactions employ natural, everyday movements of the user as a mechanism to orchestrate a device response.

Movement primitives serve as an operative framework to build interactive systems for computing with any enabling motion-sensing technology, including radar. Movement primitives consist of a movement paired with a system output; this provides a common language for designers and developers that can be utilized throughout the research and product development process. The

Figure 8.15 The Soli radar chip is embedded in the top bezel of the display (right corner).

experience designing new input languages with radar can be seen on several consumer products at Google, including the Pixel phones, displays, and thermostat devices.

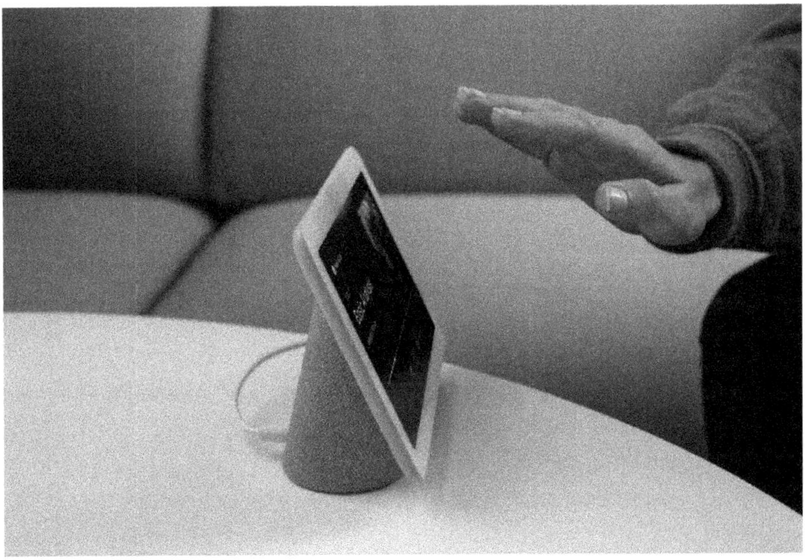

Figure 8.16 A tap of the hand over the display can pause music.

References

[1] Sutherland, I. E., "SketchPad: A Man-Machine Graphical Communication System," *AFIPS Conference Proceedings*, Vol. 23, 1963, pp. 323–328.

[2] Bolt, R. A., "'Put-That-There': Voice and Gesture at the Graphics Interface," *SIGGRAPH '80: Proceedings of the 7th Annual Conference on Computer Graphics and Interactive Techniques*, July 1980, pp. 262–270.

[3] Williams, M., and P. Green, "Interfacing the Nintendo Power Glove to a Macintosh Computer," *IVHS Technical Report-90-14*, 1990.

[4] López G., L. Quesada, and L. A. Guerrero, "Alexa vs. Siri vs. Cortana vs. Google Assistant: A Comparison of Speech-Based Natural User Interfaces," In: Nunes I. (ed.), "Advances in Human Factors and Systems Interaction," *Advances in Intelligent Systems and Computing*, Vol. 592, Springer, Cham, 2017.

[5] Zhang, X., et al., "Evaluation of Appearance-Based Methods and Implications for Gaze-Based Applications," *Proceedings of the 2019 CHI Conference on Human Factors in Computing Systems*, May 2019, pp. 1–13.

[6] Boucha, D., et al., "Controlling Electronic Devices Remotely by Voice and Brain Waves," *International Conference on Mathematics and Information Technology (ICMIT)*, 2017, pp. 38–42.

[7] Burbach, L., et al., "'Hey, Siri,' 'Ok, Google,' 'Alexa': Acceptance-Relevant Factors of Virtual Voice-Assistants," *IEEE International Professional Communication Conference (ProComm)*, 2019, pp. 101–111.

[8] Ding, I., et al., "A Kinect-Based Gesture Command Control Method for Human Action Imitations of Humanoid Robots," *International Conference on Fuzzy Theory and Its Applications (iFUZZY2014)*, 2014, pp. 208–211.

[9] Lee, M., "Development of an Open Source Platform for Brain-machine Interface: OpenBMI," *Fourth International Winter Conference on Brain-Computer Interface*, 2016, pp. 1–2.

[10] Hayashi, E., et al., "RadarNet: Efficient Gesture Recognition Technique Utilizing a Miniature Radar Sensor," *Proceedings of the 2021 CHI Conference on Human Factors in Computing Systems*, May 2021, pp. 1–14.

[11] Caine, K. E., et al., "Benefits and Privacy Concerns of a Home Equipped with a Visual Sensing System: A Perspective from Older Adults," *Proceedings of the Human Factors and Ergonomics Society Annual Meeting*, Vol. 50, Issue 2, 2006, pp. 180–184.

[12] Diraco G, et al., "Radar-Based Smart Sensor for Unobtrusive Elderly Monitoring in Ambient Assisted Living Applications," *Biosensors*, 2017.

[13] Microsoft, "Hololens: Handtracking," 2019. https://docs.microsoft.com/en-us/windows/mixed-reality/mrtk-unity/features/input/hand-tracking?view=mrtkunity-2021-05 (last accessed October 2021).

[14] Lien, J., et al., "Soli:Ubiquitous Gesture Sensing with Millimeter Wave Radar," *ACM Transactions on Graphics*, Vol. 35, Issue 4, 2016, pp. 1–19.

9

Research and Applications

This book has provided a comprehensive treatment of modern radar systems and their use for motion and presence sensing. This subject matter is becoming increasingly relevant today as technological advances, industry standardization, and a myriad of emerging applications fuel a modern renaissance for radar in the consumer space. Indeed, the unique sensing capabilities of radar in combination with their increasing scalability, manufacturability, affordability, and ease of integration suggest that radar sensors could someday become as ubiquitous as cameras, with applications throughout daily life.

In this chapter, we discuss a few of the factors driving the modern radar renaissance, their impact on radar adoption, and their outlook for the future.

9.1 Technological Trends

Several trends of technological advancements are driving research, development, and innovation in radar systems, making them increasingly useful and deployable as consumer sensors.

High-Frequency RF

There is increasing interest in radar systems operating at millimeter and submillimeter wavelengths, due to the ability to shrink the physical size of antennas and RF components as a function of wavelength. As radar operating in the 60-GHz and 77-GHz bands become commonplace today, academic and industry research and development is increasing focus at higher frequencies (e.g., in the 120-GHz band and above). At such frequencies, antenna arrays that are capable of forming highly directional beams require very small footprints and can even be integrated directly onto the radar chip package [1–5]. In addition, the potential to operate with wider bandwidths at higher frequencies allows for finer range resolution. The 122-GHz band currently has 1 GHz allocated for industrial, scientific, and medical (ISM) purposes, compared to 0.5 GHz allocated ISM in the 60-GHz band. There is potential for tens of gigahertz of unlicensed spectrum at 120 GHz, making imaging and other resolution-dependent applications increasingly attractive at such wavelengths. At terahertz or submillimeter frequencies (above 300 GHz), the intersection of RF and optical physics becomes an interesting field for exploration at such frequency bands.

Current research focuses on improving power output, signal quality, and antenna arrays for fully integrated high-frequency chipsets [1], with chipsets in the 120-GHz band beginning to be offered commercially off the shelf by silicon vendors.

Multifunctional RFICs

With the ubiquitous deployment of wireless communications chipsets (e.g., WiFi, 5G, UWB, and more), there is increasing interest and potential in leveraging these systems for radar sensing functionality [6]. Reuse of the antennas and RF front end for both communications and radar significantly increases the user value of the hardware, while introducing minimal added cost and integration effort. With such hardware reuse, UWB chipsets have the potential to support device-to-device ranging and angle of arrival, as well as device-to-user and device-to-environment radar sensing [7]. Chipsets operating with the 60-GHz 802.11ad communications standard also have attracted significant interest as platforms for enabling radar functionality [8]. There are also a lot of research activities on human sensing and biometric measurement with passive radar using WiFi signals [9,10]. The deployment of these multifunctional RFICs opens interesting opportunities for multistatic radar systems and cooperative radar networks that can leverage transmissions for both communication and radar sensing.

Deep Learning

Rapid advances in the field of deep learning are driving new techniques and algorithms for extracting meaningful information from radar data [11]. Advanced machine learning techniques are making it possible to automatically produce semantic interpretation of radar data (e.g., object and gesture recognition, scene segmentation, and user identification), paving the way for new use cases and human-machine interfaces. In addition, there is an active body of research investigating how deep learning can improve or replace traditional radar signal processing in the range, Doppler, and/or spatial domain [12].

9.2 Radar Standardization

The recent launch of the Ripple industry radar standard [13] is an important step toward greater radar hardware-software interoperability that will facilitate radar adoption and application development. Ripple seeks to standardize the API for the radar software stack, enabling radar hardware-software interoperability and accelerating the growth of innovations and applications for commodity radar, ultimately leading to better products for consumers. The Ripple technical working group has already drawn participants from a wide range of companies and industries, from silicon vendors to consumer electronics manufacturers to application developers. We expect contributions to increase from each of these sectors as the open API enables scalable participation.

9.3 Emerging Applications

The technology advancements combined with push for radar industry standardization have opened the door for rapid acceleration of new emerging applications and use cases where radar was never applied before.

Ambient Computing and IoT

The vision of ambient computing calls for contactless sensing modalities that will enable touchless interfaces with computing devices that are embedded into everyday devices and environments. Radar is a natural choice as a sensor for ambient computing due to its ability to detect users' intention, sense the environment, and provide situational awareness.

Autonomous Vehicles

Radar sensors have already been widely integrated into modern cars for enhanced safety and comfort features. Most autonomous car designers also consider radar as an integral part of the perception system. Another emerging field of radar is the in-cabin monitoring systems. As a key sensor, radar can be used to monitor children's presence to prevent them from being locked and left behind.

Noninvasive Health Sensing

As radar hardware shrinks to single chipset implementations that are easily embedded into everyday environments, radar brings the potential for contactless, continuous, effortless health monitoring. By measuring micromotions of the body surface, radar can be used to measure heartbeat, respiration rate [14, 15], and blood pressure [16–18]. Continuous measuring of these biometrics can provide significant benefits for the users. Radar plus machine learning could also transform the monitoring of glucose levels through noninvasive detecting changes of the associated dielectric properties [2, 19].

Smart Home

Besides being integrated into thermostat and doorbells, there are research activities on using radar to monitor the activities of daily living [21]. Radar has also been studied to provide fall detection of the elderly [22] as part of smart home monitoring. Foot gestures recognition such as kick has recently attracted attention as a hands-free tool [23].

References

[1] Vogelsang, F., et al., "A Highly Efficient 120 GHz and 240 GHz Signal Source in a SiGe-Technology," *IEEE BiCMOS and Compound Semiconductor Integrated Circuits and Technology Symposium*, 2020, pp. 1–4.

[2] Furqan, M., et al., "A 120-GHz Wideband FMCW Radar Demonstrator Based on a Fully Integrated SiGe Transceiver with Antenna-in-Package," *IEEE MTT-S International Conference on Microwaves for Intelligent Mobility*, 2016, pp. 1–4.

[3] Scherr, S., et al., "Miniaturized 122 GHz ISM Band FMCW Radar with Micrometer Accuracy," *European Radar Conference* (EuRAD), 2015, pp. 277–280.

[4] Girma, M. G., et al., "Miniaturized 122 GHz System-in-Package (SiP) Short Range Radar Sensor," *2013 European Radar Conference*, 2013, pp. 49–52.

[5] Ozturk E, et al., "Measuring Target Range and Velocity: Developments in Chip, Antenna, and Packaging Technologies for 60-GHz and 122-GHz Industrial Radars," *IEEE Microwave Magazine,* Vol. 18, No. 7, 2017, pp. 26–39.

[6] Liang Han, and Ke Wu, "Joint Wireless Communication and Radar Sensing Systems—State of the Art and Future Prospects," *IET Microw. Antennas Propag.*, Vol. 7, Issue 11, 2013, pp. 876–885.

[7] Saddik, G. N., et al., "Ultra-Wideband Multifunctional Communications/Radar System," *IEEE Transactions on Microwave Theory and Techniques*, Vol. 55, No. 7, 2007, pp. 1431–1437.

[8] Kumari, P., et al., "IEEE 802.11ad-Based Radar: An Approach to Joint Vehicular Communication-Radar System," *IEEE Transactions on Vehicular Technology*, Vol. 67, No. 4, 2018, pp. 3012–3027.

[9] Li, W., et al., "Passive WiFi Radar for Human Sensing Using a Stand-Alone Access Point," *IEEE Transactions on Geoscience and Remote Sensing*, Vol. 59, No. 3, 2021, pp. 1986–1998.

[10] Tang, M. C., "Vital-Sign Detection Based on a Passive WiFi Radar," *IEEE MTT-S 2015 International Microwave Workshop Series on RF and Wireless Technologies for Biomedical and Healthcare Applications*, 2015, pp. 74–75.

[11] Geng, Z., et al., "Deep-Learning for Radar: A Survey," *IEEE Access.*, 2021, p. 1-1.

[12] Brodeski, D., et al., "Deep Radar Detector," *IEEE Radar Conference (RadarConf)*, 2019.

[13] Consumer Technology Association, "Ripple," https://cta.tech/ripple (last accessed March 2022).

[14] Kocur, D., et al., "Estimation of Breathing Frequency and Heart Rate by Biometric UWB Radar," *2018 IEEE International Conference on Systems, Man, and Cybernetics (SMC)*, 2018, pp. 2570–2576.

[15] Rathna, G. N., and D. Meshineni, "Analysis of FM CW-Radar Signals to Extract Vital-Sign Information," *2021 IEEE International Conference on Electronics, Computing and Communication Technologies (CONECCT)*, 2021, pp. 1–6.

[16] Tseng, T. J., and C. H. Tseng, "Noncontact Wrist Pulse Waveform Detection Using 24-GHz Continuous-Wave Radar Sensor for Blood Pressure Estimation," *2020 IEEE/MTT-S International Microwave Symposium (IMS)*, 2020, pp. 647–650.

[17] Kawasaki, R., and A. Kajiwara, "Continuous Blood Pressure Estimation Using Millimeter Wave Radar," *2022 IEEE Radio and Wireless Symposium (RWS)*, 2022, pp. 135–137.

[18] Buxi, D., et al., "Blood Pressure Using Pulse Transmit Time from Bioimpedance and Continuous Wave Radar," *Transaction on IEEE Biomedical Engineering*, Vol. 64, No. 4, 2017, pp. 917–927.

[19] Tang, L., et al. "Non-Invasive Blood Glucose Monitoring Technology: A Review," *Sensors*, Vol. 20, No. 23, 2020.

[20] Omer, A., et al., "Blood Glucose Level Monitoring Using an FMCW Millimeter-Wave Radar Sensor," *Remote Sensing*, Vol. 12, No. 3, 2020, p. 385.

[21] Bouchard, K., et al., "Activity Recognition in Smart Homes using UWB Radars," *The 11th International Conference on Ambient Systems, Networks and Technologies (ANT)*, 2020, pp. 10–17.

[22] Amin, M., et al., "Radar Signal Processing for Elderly Fall Detection: The Future for In-Home Monitoring," *IEEE Signal Processing Magazine*, No. 33, 2016, pp. 71–80.

[23] Song, S., et al., "Foot Gesture Recognition Using High-Compression Radar Signature Image and Deep Learning," *Sensors*, Vol. 21, No. 11, 2021, p. 3937.

About the Authors

Jian Wang (SM'14) received a PhD degree in electrical engineering from the University of Victoria, Canada, and a BSc from Ocean University of China. He joined Google in 2017 as systems lead to transform Project Soli into consumer products. The Soli radar was successfully integrated in 2019 Pixel phones. Prior to Google, Dr. Wang worked for Apple on its special project of autonomous systems as a lead systems architect.

Before joining Apple, Dr. Wang worked for Rockwell Collins (now part of Raytheon) as principal systems engineer in the Advanced Radar Applications Group. There he worked on next-generation airborne multifunction radar with active electronically scanned array and novel signal processing algorithms to better detect weather and suppress ground clutter, and to detect and display vertical weather. Prior to Rockwell Collins, Dr. Wang spent more than 10 years in Raytheon and worked on both air traffic control radar and high frequency surface wave radar. There he contributed key technology enabler for the long-range radar service life extension program, which included 80 radars to provide seamless coverage of North America.

His research interests include everything related to radar, array signal processing, MIMO, detection and estimation theories, target tracking, EM simulation, and antenna design. Dr. Wang holds nine U.S./international patents, with another eight pending.

Jaime Lien is the radar research lead of Project Soli at Google Advanced Technology and Projects. She leads a technical team developing novel radar sensing techniques and systems for human perception and interaction. Soli radar technology has enabled new modes of touchless interaction in consumer wearables and devices, including Google's Pixel 4 smartphone, Home Hub, and Nest Thermostat.

Prior to Google, Dr. Lien worked as a communications engineer at NASA's Jet Propulsion Laboratory. She holds a PhD in electrical engineering from Stanford University, where her research focused on interferometric synthetic aperture radar theory and techniques. She obtained her bachelor's and master's degrees in electrical engineering from MIT. Her current research interests include radar signal processing and sensing algorithms, modeling and analysis of the underlying RF physics, and inference on radar data.

Index

Artech House Radar Library

Joseph R. Guerci, Series Editor

Adaptive Antennas and Phased Arrays for Radar and Communications, Alan J. Fenn

Adaptive Signal Processing for Radar, Ramon Nitzberg

Advanced Techniques for Digital Receivers, Phillip E. Pace

Advances in Direction-of-Arrival Estimation, Sathish Chandran, editor

Airborne Early Warning System Concepts, Maurice W. Long

Airborne Pulsed Doppler Radar, Second Edition, Guy V. Morris and Linda Harkness, editors

Aspects of Modern Radar, Eli Brookner

Aspects of Radar Signal Processing, Frank F. Kretschmer, Bernard Lewis and Wesley W. Shelton

Basic Radar Analysis, Mervin C. Budge, Jr. and Shawn R. German

Basic Radar Tracking, Mervin C. Budge, Jr. and Shawn R. German

Bayesian Multiple Target Tracking, Second Edition, Lawrence D. Stone, Roy L. Streit, Thomas L. Corwin, and Kristine L Bell

Beyond the Kalman Filter: Particle Filters for Tracking Applications, Branko Ristic, Sanjeev Arulampalam, and Neil Gordon

Cognitive Radar: The Knowledge-Aided Fully Adaptive Approach, Second Edition, Joseph R. Guerci

Coherent Radar Performance Estimation, James A. Scheer

Computer Simulation of Aerial Target Radar Scattering, Recognition, Detection, and Tracking, Yakov D. Shirman, editor

Control Engineering in Development Projects, Olis Rubin

Countermeasures for Aerial Drones, Garik Markarian and Andrew Staniforth

Deep Learning Applications of Short-Range Radars, Avik Santra, and Souvik Hazra

Deep Learning for Radar and Communications Automatic Target Recognition, Uttam K. Majumder, Erik P. Blasch, and David A. Garren

Design and Analysis of Modern Tracking Systems, Samuel Blackman and Robert Popoli

Detectability of Spread-Spectrum Signals, George Dillard

Detecting and Classifying Low Probability of Intercept Radar, Second Edition, Phillip E. Pace

FMCW Radar Design, M. Jankiraman

Fourier Transforms in Radar and Signal Processing, Second Edition, David Brandwood

Fundamentals of Short-Range FM Radar, Igor V. Komarov and Sergey M. Smolskiy

Handbook of Computer Simulation in Radio Engineering, Communications, and Radar, Sergey A. Leonov and Alexander I. Leonov

Handbook of Radar Measurement, David K. Barton

High-Resolution Radar, Second Edition, Donald R. Wehner

High Resolution Radar Cross-Section Imaging, Dean L. Mensa

Highly Integrated Low-Power Radars, Sergio Saponara, Maria Greco, Egidio Ragonese, Giuseppe Palmisano, and Bruno Neri

Introduction to LabVIEW™ FPGA for RF, Radar, and Electronic Warfare Applications, Terry Stratoudakis

An Introduction to Passive Radar, Second Edition, Hugh D. Griffiths and Christopher J. Baker

Introduction to Radar using Python and MATLAB®, Lee Andrew Harrison

Introduction to RF Equipment and System Design, Pekka Eskelinen

Linear Systems and Signals: A Primer, JC Olivier

Meter-Wave Synthetic Aperture Radar for Concealed Object Detection, Hans Hellsten

Methods and Techniques of Radar Recognition, Victor Nebabin

The Micro-Doppler Effect in Radar, Second Edition, Victor C. Chen

Microwave Radar: Imaging and Advanced Concepts, Roger J. Sullivan

Millimeter-Wave Radar Clutter, Nicholas C. Currie, Robert D. Hayes, and Robert Trebits

Millimeter-Wave Radar Targets and Clutter, Gennadiy P. Kulemin

MIMO Radar: Theory and Application, Jamie Bergin and Joseph R. Guerci

Modern Radar Systems, Second Edition, Hamish Meikle

Modern Radar System Analysis, David K. Barton

Modern Radar System Analysis Software and User's Manual, Version 3.0, David K. Barton

Monopulse Principles and Techniques, Second Edition, Samuel M. Sherman and David K. Barton

Motion and Gesture Sensing with Radar, Jian Wang and Jaime Lien

MTI and Pulsed Doppler Radar with MATLAB®, Second Edition, D. Curtis Schleher

Multifunction Array Radar, Dale Billetter

Multiple-Target Tracking with Radar Applications, Samuel S. Blackman

Multitarget-Multisensor Tracking, Applications and Advances, Volume II, Yaakov Bar-Shalom and Robert F. Popoli, editors

Multitarget-Multisensor Tracking: Applications and Advances Volume III, Yaakov Bar-Shalom and William Dale Blair, editors

Non-Line-of-Sight Radar, Brian C. Watson and Joseph R. Guerci

Photonic Aspects of Modern Radar, Edward N. Toughlian and Henry Zmuda

Practical Phased Array Antenna Systems, Eli Brookner

Practical Simulation of Radar Antennas and Radomes, Herbert L. Hirsch

Precision FMCW Short-Range Radar for Industrial Applications, Boris A. Atayants, Viacheslav M. Davydochkin, Victor V. Ezerskiy, Valery S. Parshin, and Sergey M. Smolskiy

Principles of High-Resolution Radar, August W. Rihaczek

Principles of Radar and Sonar Signal Processing, François Le Chevalier

Radar Cross-Section Analysis & Control, Asoke K. Bhattacharyya and D. L. Sengupta

Radar Cross Section, Second Edition, Eugene F. Knott, et al.

Radar Detection and Tracking Systems, Shaheen A. Hovanessian

Radar Engineer's Sourcebook, William C. Morchin

Radar Equations for Modern Radar, David K. Barton

Radar Evaluation Handbook, David K. Barton, et al.

Radar for Fully Autonomous Driving, Matt Markel, Editor

Radar Meteorology, Henri Sauvageot

Radar Range Performance Analysis, Lamont V. Blake

Radar Reflectivity of Land and Sea, Third Edition, Maurice W. Long

Radar Resolution and Complex-Image Analysis, August W. Rihaczek and Stephen J. Hershkowitz

Radar RF Circuit Design, Second Edition, Nickolas Kingsley and Joseph R. Guerci

Radar Signal Processing and Adaptive Systems, Ramon Nitzberg

Radar Signals: An Introduction to Theory and Application, Charles E. Cook

Radar System Analysis, Design, and Simulation, Eyung W. Kang

Radar System Analysis and Modeling, David K. Barton

Radar System Design and Analysis, Shaheen A. Hovanessian

Radar System Performance Modeling, Second Edition, G. Richard Curry

Radar Technology, Eli Brookner

Radar Technology Encyclopedia, David K. Barton and Sergey A. Leonov, editors

Radar Vulnerability to Jamming, Stephen Lothes

Radars, Volume 1: Monopolse Radar, David K. Barton

Radars, Volume 2: The Radar Equation, David K. Barton

Radars, Volume 3: Pulse Compression, David K. Barton

Radars, Volume 4: Radar Resolution & Multipath Effects, David K. Barton

Radars, Volume 5: Radar Clutter, David K. Barton

Radars, Volume 6: Frequency Agility and Diversity, David K. Barton

Radars, Volume 7: CW and Doppler Radar, David K. Barton

Radio Wave Propagation Fundamentals, Second Edition, Artem Saakian

Range-Doppler Radar Imaging and Motion Compensation, Jae Sok Son, et al.

Robotic Navigation and Mapping with Radar, Martin Adams, John Mullane, Ebi Jose, and Ba-Ngu Vo

Russian-English and English-Russian Dictionary of Radar and Electronics, William F. Barton and Sergey A. Leonov

Secondary Surveillance Radar, Michael C. Stevens

Signal Detection and Estimation, Second Edition, Mourad Barkat

Signal Processing in Noise Waveform Radar, Krzysztof Kulpa

Signal Processing for Passive Bistatic Radar, Mateusz Malanowski

Solid-State Radar Transmitters, Edward D. Ostroff

Space-Based Radar Handbook, Leopold J. Cantafio, Richard K. Moore, and Earl E. Swartzlander Jr.

Space-Time Adaptive Processing for Radar, Second Edition, Joseph R. Guerci

Spaceborne Weather Radar, Toshiaki Kozu and Robert Meneghini

Special Design Topics in Digital Wideband Receivers, James Tsui

Statistical Signal Characterization, Herbert L. Hirsch

Surface-Based Air Defense System Analysis, Robert H. MacFadzean

Synthetic Array and Imaging Radars, Shaheen A. Hovanessian

Systems Engineering of Phased Arrays, Rick Sturdivant, Clifton Quan, and Enson Chang

Techniques of Radar Reflectivity Measurement, Nicholas C. Currie

Theory and Practice of Radar Target Identification, August W. Rihaczek and Stephen J. Hershkowitz

Time-Frequency Signal Analysis with Applications, Ljubiša Stanković, Miloš Daković, and Thayananthan Thayaparan

Time-Frequency Transforms for Radar Imaging and Signal Analysis, Victor C. Chen and Hao Ling

Transmit Receive Modules for Radar and Communication Systems, Rick Sturdivant and Mike Harris

For further information on these and other Artech House titles, including previously considered out-of-print books now available through our In-Print-Forever® (IPF®)
program, contact:

Artech House
685 Canton Street
Norwood, MA 02062
Phone: 781-769-9750
Fax: 781-769-6334
e-mail: artech@artechhouse.com

Artech House
16 Sussex Street
London SW1V HRW UK
Phone: +44 (0)20 7596-8750
Fax: +44 (0)20 7630-0166
e-mail: artech-uk@artechhouse.com

Find us on the World Wide Web at: www.artechhouse.com